Introduction to Solar Cells

Introduction to Solar Cells

Ammar Nayfeh
Khalifa University, Abu Dhabi, United Arab Emirates

Sabina Abdul Hadi
University of Dubai, Dubai, United Arab Emirates

WILEY

This edition first published 2026

Registered Office
John Wiley & Sons, Inc., 111 River Street, Hoboken, NJ 07030, USA
John Wiley & Sons Ltd New Era House, 8 Oldlands Way, Bognor Regis, West Sussex, PO22 9NQ, UK

For details of our global editorial offices, customer services, and more information about Wiley products visit us at www.wiley.com.

The manufacturer's authorized representative according to the EU General Product Safety Regulation is Wiley-VCH GmbH, Boschstr. 12, 69469 Weinheim, Germany, e-mail: Product_Safety@wiley.com.

Library of Congress Cataloging-in-Publication Data applied for:

Hardback ISBN : 9781394197392

Cover Design: Wiley
Cover Image: Scanrail/stock.adobe.com

Set in 9.5/12.5pt STIXTwoText by Straive, Chennai, India
Printed and bound by CPI Group (UK) Ltd, Croydon, CR0 4YY

C9781394197392_170726

Contents

About the Authors

Professor Ammar Nayfeh is an internationally recognized expert in nanotechnology, advanced semiconductor devices, and photovoltaic materials, with over two decades of experience in academia and industry. He received his bachelor's degree in Electrical Engineering from the University of Illinois Urbana-Champaign in 2001, and his master's and Ph.D. degrees from Stanford University in 2003 and 2006, respectively, under the supervision of Professor Krishna Saraswat. His doctoral research focused on the heteroepitaxy of germanium on silicon for advanced electronic and photonic devices.

Following his Ph.D., he joined Advanced Micro Devices (AMD) in collaboration with IBM, and later became part of the Silicon Valley startup Innovative Silicon (ISi). He also served as a part-time professor at San José State University. In 2010, he joined MIT as a visiting scholar before becoming a faculty member at Khalifa University in the United Arab Emirates. He has also served as a visiting professor at Stanford University.

Professor Nayfeh is currently an Associate Professor in the Department of Electrical Engineering at Khalifa University and serves as the Electronics Lead of the Research & Innovation Center for Graphene and 2D Materials (RIC2D). His research includes the development of novel materials and nanostructures for next-generation photovoltaic devices, with a focus on improving efficiency and enabling low-cost, high-performance solar energy technologies.

He has authored or co-authored over 170 scientific publications, holds 3 patents, and has co-authored 3 books: *Silicon–Germanium (SiGe) Alloys for Photovoltaic Applications*, *Nanomaterials-Based Charge Trapping Memory Devices*, and *Integrated Silicon-Metal Systems at the Nanoscale: Applications in Photonics, Quantum Computing, Networking, and Internet*. He also serves as an editor for the *IEEE Journal of Photovoltaics* (JPV) and *Scientific Reports* (Nature Portfolio) and part of the organizing committee for the IEEE PVSC.

Professor Nayfeh has extensive teaching experience in semiconductor devices and renewable energy technologies. He regularly teaches courses including *Introduction to Solar Cells* and *Physics of Solar Cells* at both undergraduate and Ph.D. levels, mentoring students in the theoretical foundations and practical design of photovoltaic systems.

Professor Nayfeh is a Senior Member of IEEE, a member of Materials Research Society (MRS) and The Electrochemical Society (ECS), the Stanford Alumni Association, and the Mohammed Bin Rashid Academy of Science (MBRAS). Professor Nayfeh continues to focus his research on nanotechnology solutions for next-generation low-power electronic, photonic, and solar energy systems. Through this textbook, he aims to provide students and researchers with a rigorous yet accessible foundation in solar cell physics and technology.

Sabina Abdul Hadi is currently an Associate Professor of Electrical Engineering in the College of Engineering and Information Technology at the University of Dubai. She received her Ph.D. degree in Interdisciplinary Engineering from Masdar Institute of Science and Technology, Abu Dhabi, United Arab Emirates, in 2016. Her doctoral research focused on the modeling, fabrication, and experimental characterization of III–V/Si multi-junction solar cells.

From 2016 to 2018, Dr. Sabina Abdul Hadi served as a postdoctoral fellow at Khalifa University, where she worked on advanced photovoltaic technologies and memristor-based sensing applications. She joined the University of Dubai in 2018 and continues to teach and conduct research in semiconductor device physics and emerging electronic materials.

Dr. Sabina Abdul Hadi has co-authored more than 40 peer-reviewed journal articles and conference proceedings and is a co-author of the book *Silicon–Germanium (SiGe) Alloys for Photovoltaic Applications*. Her research interests include semiconductor physics, device modeling and fabrication, nanotechnology, memristor devices and their applications, thin-film materials, anti-reflective coating design, emerging photovoltaic technologies, semiconductor device characterization, and the application of artificial intelligence and data-driven methods in semiconductor and photovoltaic technologies.

She is actively involved in the IEEE community and currently serves as Vice-Chair of the IEEE Circuits and Systems Society (CASS), Solid-State Circuits Society (SSCS), and Power Electronics Society (PELS), UAE Section. She previously served as Chair of the IEEE Power and Energy Society (PES), UAE Section from 2020 to 2022. Dr. Sabina Abdul Hadi has contributed as a technical and organizing committee member for several IEEE conferences over the years.

Preface

This textbook is an introduction to solar cells aimed at junior- or senior-level undergraduate students. It may also serve as a reference for M.Sc. and Ph.D. students working in solar and photovoltaic research, as well as for solar device engineers who wish to review the fundamental principles underlying their professional practice. This book provides the background needed to understand both the motivation for using solar energy and the physics of solar cell operation, from photon absorption to electron transport, along with device design and testing.

The text covers the physics of light, semiconductors, junctions, solar cell design, and measurement techniques. Detailed in-chapter exercises with fully worked solutions are included to help students master the material. End-of-chapter exercises are also provided for additional practice and deeper exploration. The guiding philosophy of this book is to emphasize a conceptual and physics-based understanding of solar cells, supported by mathematical problem-solving and engineering insight. We believe this approach provides students with the foundational knowledge required to analyze, design, and innovate in solar cell technologies.

Goals of the Book

With the current challenges of climate change driven by fossil fuel consumption and the growing need for energy independence, it is essential that undergraduate students learn the fundamentals of solar energy and solar cell technology. Solar energy plays a central role in the global transition toward renewable energy sources. This textbook has the following goals:

- Provide a comprehensive foundation in solar cell physics and technology.
- Equip students with the skills required for solar cell design and performance optimization.

- Inspire and prepare students for advanced study and careers in solar and renewable energy.
- Develop future leaders capable of contributing to the global energy transition driven by solar power.

The book also supports students who plan to enter the workforce immediately after completing their undergraduate studies by providing them with a solid understanding of solar energy in the rapidly growing renewable energy industry. For instructors, this text serves as a comprehensive resource for an introductory course on solar cells for junior and senior undergraduate students, complete with examples and exercises that support effective teaching. Ultimately, we hope this book can serve as an initial spark that motivates students to pursue careers in photovoltaics and clean energy disciplines that are critical for the future global workforce.

Book Format and Structure

The book is divided into seven chapters. Chapter 1 introduces the motivation behind solar energy, including climate change and the global energy transition. Chapter 2 covers the fundamental physics of light required to understand solar cell operation. Chapter 3 provides a comprehensive overview of semiconductor physics, establishing a strong foundation for understanding why semiconductors are used in solar cells. Chapter 4 discusses the physics of semiconductor *pn* junctions as the key building block of solar cells. Chapter 5 presents a detailed explanation of solar cell operation, including the derivation and interpretation of key performance parameters. Chapter 6 focuses on solar cell design and strategies for improving efficiency. Finally, Chapter 7 provides an overview of novel efficiency enhancement methods.

Each chapter includes in-chapter exercises with detailed solutions, chapter summaries, and end-of-chapter exercises consisting of multiple-choice questions, short conceptual problems, and quantitative problem-solving exercises. This combination of theory, worked examples, summaries, and exercises is designed to provide students with a strong foundation in solar cell physics and design.

In addition to the core chapters, several supplementary features are included. A glossary of key terms is provided to support effective studying, along with a comprehensive list of symbols and definitions to guide students and facilitate reading. An appendix is also included and consists of three sections: Section A contains a complete formula sheet with page references; Section B provides tables of important constants, unit conversions, and semiconductor parameters; and Section C includes key charts and graphs.

We also provide online resources for both students and instructors. For students, chapter examples and end-of-chapter summaries are available for download in PDF format, along with an equation sheet. QR codes placed throughout the book link to additional online resources. For instructors, draft presentation slides in PowerPoint format are provided to facilitate lecture preparation, along with solutions to the end-of-chapter exercises.

Acknowledgments

The idea for this book originated in 2022, driven by the growing need to educate students in the physics of solar cells for a rapidly evolving workforce shaped by the global energy transition. During this time, Professor Nayfeh developed a new course at Khalifa University entitled *Fundamentals of Solar Cells*, further highlighting the need for a dedicated textbook. Professor Sabina Abdul Hadi's experience teaching undergraduate electronics courses at the University of Dubai strongly influenced many of the worked examples, exercises, and figures throughout this book. We would like to acknowledge Khalifa University and the University of Dubai for their support in completing this textbook. We would also like to thank our colleagues and international collaborators who have inspired us throughout our journey in solar and photovoltaic research and education.

Last but not least, we would like to thank our families for their constant patience, support, and encouragement throughout the writing process. Without their unwavering support and sacrifices, this textbook would not have been possible.

Finally, we hope this book serves as the beginning of a long and rewarding career path for students pursuing solar cell research and innovation, fostering a lasting passion for renewable and green energy, and empowering the next generation of engineers and scientists to develop technologies that safeguard our planet and drive the global transition toward sustainable energy.

Symbols and Abbreviations

Symbols

A	Device area; junction cross-sectional area	(cm^2)
$A(\lambda)$	Absorption fraction	
B	Material constant in intrinsic concentration	(cm^{-3} $K^{-3}/^2$)
c	Speed of light in medium	(m/s)
c_0	Speed of light in vacuum	(m/s)
d	Physical thickness	
d_i	Thickness of layer i	
E	Electric field	(V/cm)
E	Photon energy, $E = h\nu$	(eV)
E(x)	Electric field as function of position	(V/cm)
E_g	Semiconductor bandgap energy	(eV or J)
E_0	Peak electric field	(V/cm)
E_c	Conduction band energy	(eV)
E_d	Donor energy level	(eV)
E_f	Fermi level	(eV)
E_i	Intrinsic Fermi level	(eV)
EQE	External quantum efficiency	
E_v	Valence band energy	(eV)
$f_F(E)$	Fermi–Dirac distribution function	
G	Generation rate	(cm^{-3} s^{-1})
FF	Fill factor	
g_{op}	Optical generation rate	(cm^{-3} s^{-1})

h	Planck constant	(J·s)
I	Current	(A)
I_0	Reverse saturation current	(A)
I_d	Dark diode current	(A)
I_m	Current at maximum power point	(A)
I_p	Photogenerated current	(A)
IQE	Internal quantum efficiency	
I_r	Reflected optical power	(W)
I_{SC}	Short-circuit current	(A)
I_t	Total current	(A)
J	Current density	(A/cm^2)
J_{ph}	Photogenerated current density	(A/cm^2)
J_{sc}	Short-circuit current density	(A/cm^2)
J_1	Current density of top subcell	(A/cm^2)
J_2	Current density of bottom subcell	(A/cm^2)
J_i	Current density of subcell i	(A/cm^2)
J_m	Current density at maximum power	(A/cm^2)
J_n	Electron current density	(A/cm^2)
J_p	Hole current density	(A/cm^2)
J_{ph}	Photogenerated current density	(A/cm^2)
J_{sc}	Short-circuit current density	(A/cm^2)
J_t	Tandem current density	(A/cm^2)
J_{total}	Total current density	(A/cm^2)
K	Kelvin	
k	Boltzmann constant	(eV/K)
L	Absorption length ($1/\alpha$)	
L_n	Electron diffusion length	
L_p	Hole diffusion length	
L_{eff}	Effective optical path length	
L_{opt}	Optical path length	
L_p	Hole diffusion length	
M	Transfer matrix	
m^*	Effective mass	(kg)
m_n^*	Electron effective mass	(kg)
m_p^*	Hole effective mass	(kg)
n	Electron concentration	(cm^{-3})

$\tilde{n}_x$	Complex refractive index	unitless
$N(E)$	Density of states function	cm^{-3} eV^{-1}
n_0	Equilibrium electron concentration	(cm^{-3})
N_a	Acceptor concentration	(cm^{-3})
N_a^-	Ionized acceptor concentration	(cm^{-3})
n_{air}	Refractive index of air	unitless
N_c	Effective density of states (conduction band)	(cm^{-3})
N_d	Donor concentration	(cm^{-3})
N_d^+	Ionized donor concentration	(cm^{-3})
n_p	Minority electron concentration on p-side	(cm^{-3})
n_s	Refractive index of semiconductor	unitless
n_{Si}	Refractive index of silicon	unitless
N_v	Effective density of states (valence band)	(cm^{-3})
p	Hole concentration	(cm^{-3})
P	Electrical power output ($P = IV$)	(W)
p_n	Minority hole concentration on n-side	(cm^{-3})
p_0	Equilibrium hole concentration	(cm^{-3})
P_{in}	Incident solar power density	(W/cm^2)
P_{max}	Maximum output power	(W)
P_{out}	Output power	(W)
q	Elementary charge	(C)
QE	Quantum efficiency	
Q_n^+	Total positive charge on n-side	(C)
Q_p^-	Total negative charge on p-side	(C)
R	Resistance	(Ω)
$\mathscr{R}$	Recombination rate	(cm^{-3} s^{-1})
$R(\lambda)$	Reflectance	
r_{12}	Amplitude reflection coefficient	
$S_\lambda(\lambda)$	Spectral irradiance	(W m^{-2} nm^{-1})
$S_i(E)$	Spectral response of subcell i	
$\mathscr{T}(\lambda)$	Transmittance	
T	Temperature	(K)
t_{12}	Amplitude transmission coefficient	
v	Carrier velocity	(cm/s)
V	Electrostatic potential	(V)

V_0	Built-in potential	(V)
V_a	Applied voltage	(V)
V_{FB}	Forward bias barrier	(V)
V_m	Voltage at maximum power point	(V)
V_{oc}	Open-circuit voltage	(V)
V_{oc1}	Open-circuit voltage of top cell	(V)
V_{oc2}	Open-circuit voltage of bottom cell	(V)
V_{RB}	Reverse bias barrier	(V)
V_{tandem}	Total open-circuit voltage	(V)
$\overline{v}_x$	Average drift velocity	(cm/s)
W	Depletion width	(cm)
X	Concentration factor	
x_n	Depletion width on n-side	(cm)
x_p	Depletion width on p-side	(cm)
α	Absorption coefficient	(cm^{-1})
δ	Absorption depth ($1/\alpha$)	(cm)
Δn	Excess electron concentration	(cm^{-3})
δn	Steady-state excess electron concentration	(cm^{-3})
Δp	Excess hole concentration	(cm^{-3})
$\Delta \lambda$	Wavelength interval	(nm)
η	Solar cell conversion efficiency	
θ	Angle of incidence	degrees or radians
κ	Extinction coefficient	
λ	Wavelength	(nm)
μ_n	Electron mobility	(cm^2/V·s)
μ_p	Hole mobility	(cm^2/V·s)
μ_{total}	Total carrier mobility	(cm^2/V·s)
ν	Frequency	(Hz)
σ_n	Electron conductivity	$(\Omega\ cm)^{-1}$
σ_p	Hole conductivity	$(\Omega\ cm)^{-1}$
τ_n	Electron minority carrier lifetime	(s)
τ_p	Hole minority carrier lifetime	(s)
$\Phi(E)$	Spectral photon flux as function of energy	(photons · $cm^{-2}\ s^{-1}\ eV^{-1}$)
$\Phi(\lambda)$	Spectral photon flux as a function of wavelength	(photons · $cm^{-2}\ s^{-1}\ nm^{-1}$)
φ_i	Phase thickness of layer i	(radians)

$\mathbb{I}(x)$	Optical intensity at depth x	(W/cm^2)
$\mathbb{I}_0$	Incident optical intensity	(W/cm^2)
$\mathbb{I}_t$	Transmitted optical power	(W)

Abbreviations

2J	Two-junction solar cell
2T	Two-terminal configuration
AI	Artificial intelligence
AM	Air Mass
AM 0	Extraterrestrial solar spectrum
AM 1.5D	Air Mass direct terrestrial solar spectrum
AM 1.5G	Air Mass global terrestrial solar spectrum
ARC	Anti-reflective coating
BIPV	Building-integrated photovoltaics
CdTe	Cadmium telluride
CH_4	Methane
CIGS	Copper indium gallium selenide
CO_2	Carbon dioxide
CPV	Concentrated photovoltaics
DSSCs	Dye-sensitized solar cells
eV	Electron volt
EHP	Electron–hole pair
EQE	External quantum efficiency
FF	Fill factor
GaAs	Gallium arsenide
Ge	Germanium
HFCs	Hydrofluorocarbons
HJT	Heterojunction technology
IEA	International Energy Agency
III–V	Group III–V semiconductors
InGaP	Indium gallium phosphide
IoT	Internet of Things
IPCC	Intergovernmental Panel on Climate Change
IQE	Internal quantum efficiency
IR	Infrared radiation

IRENA	International Renewable Energy Agency
IV	Current–voltage characteristic
kWh	Kilowatt-hour
LED	Light-emitting diode
MPP	Maximum power point
MW	Megawatt
N_2O	Nitrous oxide
NF_3	Nitrogen trifluoride
NLR	National Laboratories of the Rockies (previously NREL: National Renewable Energy Laboratory)
OPL	Optical path length
OPVs	Organic photovoltaics
OSC	Organic solar cell
PCE	Power conversion efficiency
PERC	Passivated emitter rear contact
PFCs	Perfluorocarbons
ppm	Parts per million
PV	Photovoltaic
QD	Quantum dot
SF_6	Sulfur hexafluoride
Si	Silicon
Si_3N_4	Silicon nitride
SQ limit	Shockley–Queisser limit
TCO	Transparent conductive oxide
TiO_2	Titanium dioxide
TMM	Transfer matrix method
TOPCon	Tunnel oxide passivated contact
UAE	United Arab Emirates
WRI	World Resources Institute
WWA	World Weather Attribution

About the Companion Website

This book is accompanied by a companion website:

www.wiley.com/go/introsolarcells

To access the website materials, click on the link under Related Resources on the book's page on wiley.com. You can also find this by searching www.wiley.com for the ISBN 9781394197392.

The website includes:

- Test Bank
- Problems and Solutions
- PPT Slides
- Book-Examples
- Chapter Solutions Selected-Students
- Chapter Summaries

1

Introduction and Motivation

1.1 About the Chapter

This chapter provides the scientific and societal motivation for the study of solar energy and photovoltaic technologies. It begins by explaining the greenhouse effect and its role in global warming, followed by an overview of the major sources of greenhouse gas emissions and their impact on climate change. The chapter then introduces solar energy as a clean and sustainable alternative, highlighting its rapid technological progress, declining costs, and growing global deployment. Issues of energy independence, rising energy demand, and the environmental advantages and limitations of solar power are also discussed. Finally, a brief historical background and an overview of current solar cell technologies are presented to set the foundation for the detailed physical principles explored in the subsequent chapters.

Key Topics

- Greenhouse effect/climate change
- Environmental/societal impacts
- Solar energy/photovoltaic effect
- Cost, scalability, energy independence
- History of solar cell technologies

1.2 The Greenhouse Effect

When sunlight reaches Earth, some of the light is reflected in the atmosphere by ice and clouds, while the remaining light is absorbed by Earth. Specifically for Earth, 30% of light is reflected while 70% is absorbed. Reflected sunlight that is bounced back into space does not contribute to warming of Earth. However, some of the sunlight absorbed by Earth is later re-emitted as infrared (IR) radiation, which is partially trapped by greenhouse gases in the atmosphere and contributes to warming. IR is a type of electromagnetic radiation that falls between visible light and microwaves on the electromagnetic spectrum. It is invisible to the human eye but can be felt as heat ***(physics of light is covered in Chapter 2).*** Essentially, all objects with a temperature above absolute zero emit IR. If the IR is unable to escape, it can be trapped in the atmosphere, and this causes the Earth to increase in temperature. This phenomenon is called the *greenhouse effect*. The name is coined because it describes the effect where certain gases in the atmosphere trap heat, like how the glass of a greenhouse traps heat inside, keeping the interior warmer than the outside. The greenhouse gases released into the atmosphere by human activity are mostly from the burning of fossil fuels, but other major sources include deforestation, agriculture, and industrial processes. Examples of greenhouse gases are carbon dioxide (CO_2), methane (CH_4), nitrous oxide (N_2O); industrial gases such as hydrofluorocarbons (HFCs), perfluorocarbons (PFCs), sulfur hexafluoride (SF_6), nitrogen trifluoride (NF_3); and water vapor. Figure 1.1 shows this effect in visual form.

CO_2 makes up the majority of greenhouse gas emissions, while smaller amounts of CH_4 and N_2O are also emitted. These gases are released primarily during the combustion of fossil fuels, such as coal, oil, and natural gas, for electricity generation and other energy uses. In 2022, the largest sources of greenhouse gas emissions globally are tied to key economic sectors. The energy sector remains the dominant contributor, accounting for approximately 75.7% of emissions through electricity generation, heating, and transportation due to the burning of fossil fuels like coal, oil, and natural gas [1]. The agriculture sector contributes about ~12% of greenhouse gas emissions, largely due to CH_4 from livestock and rice cultivation, and N_2O from fertilizer use. Industrial activities, including cement

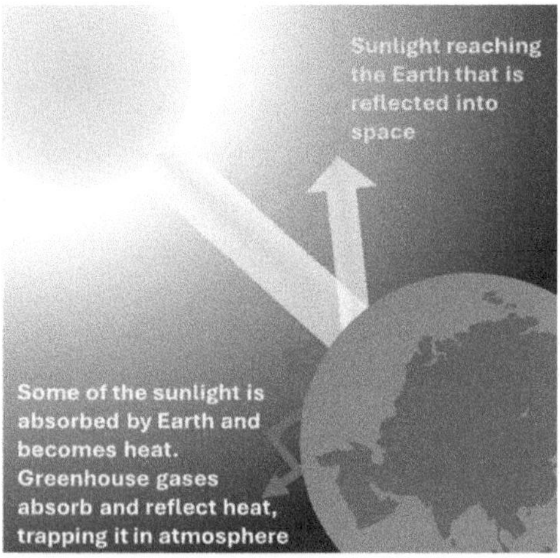

Figure 1.1 Illustration of sunlight being reflected into space and some of it being trapped in Earth's atmosphere by greenhouse gases.

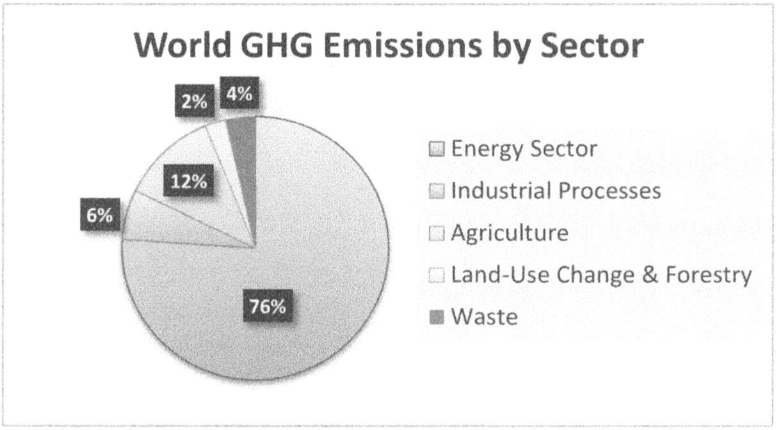

Figure 1.2 World greenhouse gas emissions by sector in 2022. *Source:* [1] / Climate Watch.

and steel production, are responsible for around 6.3% of emissions through both chemical processes and energy consumption. Waste management systems, such as landfills and wastewater treatment facilities, contribute approximately 4% of emissions, primarily in the form of CH_4 [1]. Finally, changes in land use such as deforestation account for about 2% of emissions by reducing the Earth's natural capacity to absorb CO_2. Moreover, understanding the role of each sector and their respective contributions is essential for developing effective climate mitigation strategies (Figure 1.2).

Transitioning to clean energy sources is essential because the energy sector alone contributes approximately 76% of global greenhouse gas emissions. By reducing our reliance on fossil fuels and investing in renewable technologies, we can significantly cut emissions and mitigate the impacts of climate change. As of mid-2025, the concentration of CO_2 in Earth's atmosphere has reached approximately 426 parts per million (ppm) [2]. Unfortunately, this marks a continued rise from previous years and reflects the highest levels since CO_2 started to be recorded. Historically, CO_2 concentrations have been below 300 ppm, measured from preserved air samples in ice cores [3]. Figure 1.3 illustrates the historical trend of global CO_2 emissions from 1990 to 2022, based on data from Climate Watch (World Resources Institute) [4].

1.3 Climate Change

The increase in greenhouse effect has already raised global average temperatures by approximately 1.1°C since pre-industrial times [5]. The negative effect of this

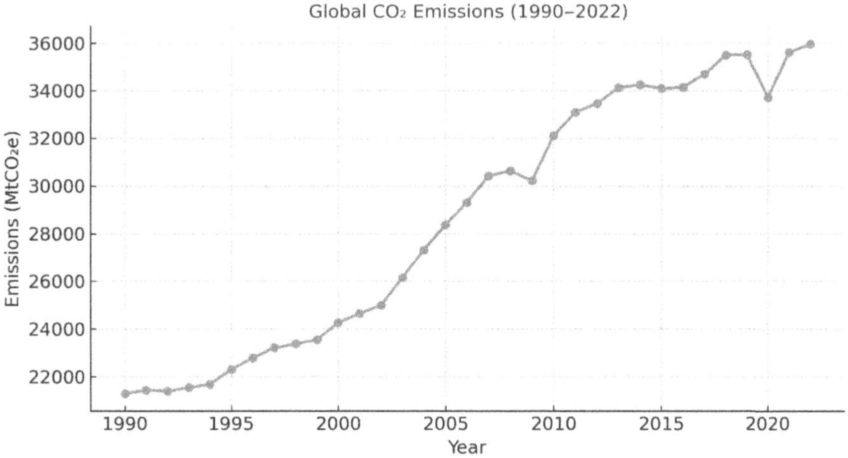

Figure 1.3 Global CO_2 emissions from 1990 to 2022. *Source:* [4] / Climate Watch.

increase is already evident. One of the most alarming is the rise in global sea levels, driven by melting glaciers and polar ice caps, which has led to increased coastal flooding and erosion in vulnerable regions. For instance, low-lying island nations like the Maldives are facing existential threats due to encroaching seas. Simultaneously, coral reef bleaching caused by rising ocean temperature has devastated marine ecosystems, particularly in the Great Barrier Reef, where repeated bleaching events have killed large swaths of coral [6, 7]. On land, record-breaking heatwaves have scorched cities across Europe and North America, straining power grids and endangering public health. For example, the Seattle heat wave in June 2021 saw temperatures reach 42.2°C, a record-breaking event in a region where such extremes are historically rare. Scientists attribute its severity to climate change, which is increasing the frequency and intensity of heat domes trapping hot air over areas like the Pacific Northwest that are typically temperate. Additionally, prolonged droughts in regions such as the Horn of Africa and the American Southwest have severely disrupted agriculture and water supplies, while increasingly intense hurricanes and floods have caused billions of dollars in damage and displaced countless communities worldwide. For example, during April 2024, the United Arab Emirates experienced its most extreme rainfall event on record, with some areas receiving nearly 300 mm of rain in a single day, more than the typical annual average [8]. This rare amount caused widespread flooding, disrupting transportation and infrastructure across cities like Dubai and Sharjah. According to World Weather Attribution group, climate change made this event 10–40% more intense by increasing atmospheric moisture and amplifying natural climate variability such as El Niño [9]. These findings highlight how global warming is making extreme weather events more frequent and severe, even in typically arid regions like the UAE. These examples underscore the urgent need for global action to mitigate further harm and to continue pushing for the transition to clean and renewable energy.

1.4 Solar Energy

Solar energy is a renewable and sustainable form of energy from the Sun's radiation. It can be harnessed using technologies such as photovoltaic (PV) solar cells, which convert sunlight directly into electricity, and solar thermal systems, which use sunlight to generate heat. Every hour, the Sun delivers about 430 quintillion joules (4.3×10^{17} J) of energy to Earth, which is more than the total energy consumed by humanity in an entire year. The most widely used method to harness solar energy today is by PV solar cells, where semiconductor materials, like silicon, absorb photons from sunlight and release electrons, creating an electric current. This process, known as the PV effect, enables solar panels to generate clean electricity without combustion or emissions. This was all made possible with advances in quantum mechanics, combined with semiconductor device physics and fabrication. The physics of PV solar cells will be the focus of this book. Most importantly, since solar energy does not rely on fossil fuels, it produces *zero greenhouse gases* during operation, making it a vital tool in the global effort to combat climate change.

The main advantage of solar energy in addressing climate change lies in its ability to replace carbon-intensive energy sources. Solar energy offers several advantages that make it an ideal energy solution. It is abundant and accessible, with Earth receiving more energy from the Sun in one hour than the world uses in an entire year. It is scalable, capable of powering everything from small homes to large industrial facilities. It is increasingly affordable, with the cost of solar PV systems having dropped by more than 90% since 2010 [10]. Table 1.1 compares how seven major renewable-power technologies performed in 2010 and 2023 in terms of total installed cost (US$/kW), capacity factor (%), and levelized cost of electricity (US$/kWh), highlighting changes over this period [10].

In fact, solar PV is now one of the cheapest sources of new electricity generation globally, thanks to technological advancements, increased production scale, and global policy support. These advancements position solar PV as one of the most affordable and sustainable energy sources available today. Figure 1.4 shows economies of scale of the PV industry, where the price of a PV panel has been consistently reducing with increased total PV installed capacity, globally. The figure is adapted from [11].

Table 1.1 Total installed cost and levelized cost of electricity by technology (year 2010 vs. 2023).

	Total installed costs (2023 US$/kW)			Levelized cost of electricity (2023 US$/kWh)		
Technology	**2010**	**2023**	**Percent change**	**2010**	**2023**	**Percent change**
Bioenergy	3010	2730	−9	0.084	0.072	−14
Geothermal	3011	4589	52	0.054	0.071	31
Hydropower	1459	2806	92	0.043	0.057	33
Solar PV	5310	758	−86	0.46	0.044	−90
CSP	10 453	6589	−37	0.393	0.117	−70
Onshore wind	2272	1160	−49	0.111	0.033	−70
Offshore wind	5409	2800	−48	0.203	0.075	−63

Source: [10] / IRENA – International Renewable Energy Agency.

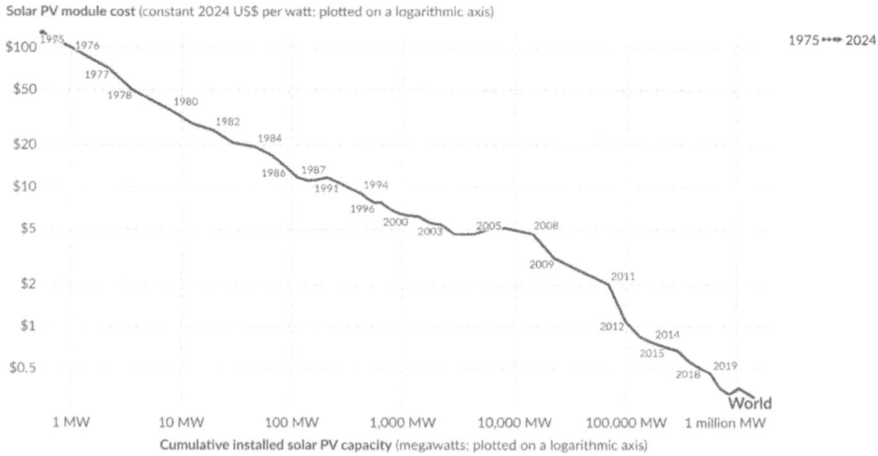

Figure 1.4 Solar PV panel prices (2024 US$) vs. global cumulative installed PV capacity (MW). *Source:* [11] / Our World in Data / CC BY 4.0.

In addition, solar energy is accessible, enabling energy independence and electrification in remote or underserved regions. These characteristics make solar energy not only environmentally beneficial but also economically and socially transformative. As of 2025, solar energy is the fastest-growing energy source in the world [12]. According to the International Energy Agency (IEA), over 270 terawatt-hours (TWh) of new electricity generation in 2022 alone of new solar were added, and its deployment continues to accelerate [13]. China leads in total installed capacity, followed by the United States, India, and several European nations. Developing countries are also embracing solar to expand access to electricity and reduce reliance on diesel generators. The IEA has identified solar as the only energy technology currently on track to meet the global climate goals, projecting that it could become the largest source of electricity worldwide before 2030 [14]. If fully deployed, solar energy could prevent billions of tons of CO_2 emissions by mid-century, significantly slowing the pace of climate change. Solar energy represents a powerful and practical solution to the climate crisis. Its clean and renewable nature allows it to reduce greenhouse gas emissions, mitigate climate change effects, and support sustainable development. As technology advances and costs continue to fall, solar energy will play an increasingly central role in the global energy transition. Moreover, investing in solar infrastructure and policy support today is essential for securing a livable climate and a resilient energy future.

1.5 Energy Independence

In addition to helping mitigate climate change, solar energy is a powerful driver of energy independence and a solution to growing global energy demands. Currently, many countries rely heavily on imported fossil fuels, which subjects them to geopolitical tensions, price fluctuations, and supply chain disruptions. By investing in solar infrastructure, nations can reduce their dependence on foreign energy sources, strengthening their energy security, thus stabilizing their industry and economy. Because solar energy is locally generated and widely available, it is especially valuable for regions seeking to stabilize their energy supply and reduce vulnerability to external factors. Furthermore, global energy consumption continues to rise due to population growth, urbanization, and expanding industrial activity. With the rise of Internet of Things (IoT) and heavy reliance of industry on artificial intelligence (AI), computational energy needs are also increasing exponentially. Figure 1.5 shows global energy consumption (TWh) by source over the years (reproduced from Our World in Data [15]). With an exponential increase in energy demand, fossil fuels remain the dominant source of energy globally, as of 2024.

Meeting this increasing demand sustainably is a major challenge, and solar energy offers a scalable and clean alternative. Figure 1.6 shows electricity generation (kWh) per capita by source in 2024, showing top producing countries. Sweden and Canada have the highest electricity generation per capita, and most of their electricity is generated from renewable and nuclear sources of energy. However, it is clear that solar PV generation is not as widely deployed as other

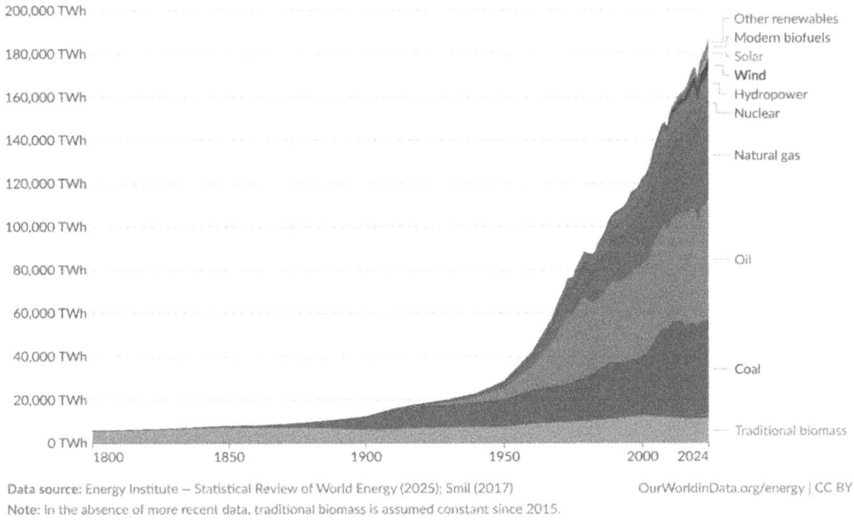

Figure 1.5 Global energy consumption (TWh) by source over the years. *Source:* [15] Our World in Data / CC BY 4.0.

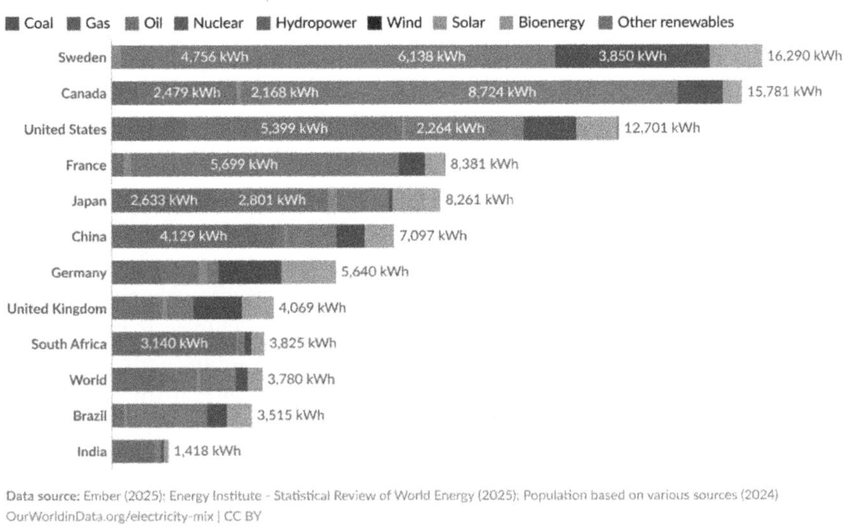

Figure 1.6 Electricity generation (kWh) per capita by source in 2024, showing top producing countries. Other renewables include geothermal, tidal, and wave generation (reproduced from Our World in Data: Ember (2025); Energy Institute—Statistical Review of World Energy (2025); population based on various sources (2024)—with major processing by Our World in Data). *Source:* [15] / Our World in Data / CC BY 4.0.

energy sources, peaking at about 800–850 kWh in the United States, Germany, and Japan. However, the ability of PV systems to be deployed at various scales, from stand-alone rooftop panels to utility-scale solar farms connected to the grid, makes them adaptable to the diverse and growing energy needs. As solar technology becomes more efficient and affordable, it not only supports climate goals but also empowers communities and nations to take control of their energy future.

Despite its many benefits, solar energy is not without criticism. Some environmental concerns have been raised regarding the land use required for large solar farms, which can disrupt local ecosystems and biodiversity [16]. Additionally, the manufacturing and disposal of solar panels involve materials and processes that can generate waste and pollution if not properly managed. However, these drawbacks are minor when compared to the relatively long (~25 years) and almost zero-emissions operational life of solar panels and long-term environmental damage caused by fossil fuel combustion. The emissions, air pollution, and climate impacts associated with coal, oil, and gas far outweigh the localized and manageable challenges of solar energy. With proper planning, recycling programs, and sustainable practices, the negative impacts of solar can be minimized, making it one of the most promising and responsible energy solutions available today. Solar is the fastest-growing energy sector today, and it is predicted to overcome all other energy sources by year 2040 [17], as shown in Figure 1.7.

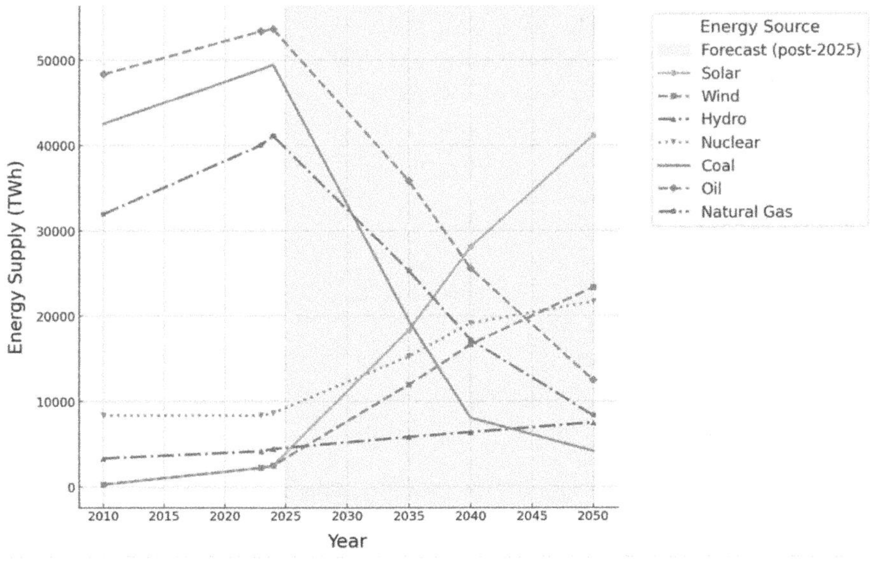

Figure 1.7 Global energy supply trends for 2010–2050 in TWh. *Source:* [17] / International Energy Agency (IEA) / CC BY 4.0.

1.6 Brief History of Solar Energy and Photovoltaics

The history of solar energy stretches from ancient civilizations to today's global energy transition. Early societies like the Greeks and Romans used passive solar heating in architecture, but the scientific journey began in 1839 when Alexandre Edmond Becquerel discovered the PV effect, which is the generation of electricity from light [18]. In 1873, Willoughby Smith observed photoconductivity in selenium, and by 1883, Charles Fritts built the first solar cell using selenium and gold, though it had less than 1% efficiency [19]. A major theoretical breakthrough came in 1905, when Albert Einstein explained the photoelectric effect, earning the 1921 Nobel Prize in Physics, laying the quantum foundation for PVs [20, 21]. In 1954, Bell Labs scientists Daryl Chapin, Calvin Fuller, and Gerald Pearson created the first practical silicon solar cell with 6% efficiency, launching the modern era of PVs. Solar cells powered satellites like Vanguard I in 1958, and the oil crises of the 1970s spurred terrestrial applications. In 1985, Australian researchers Martin Green and Andrew Blakers at the University of South Wales achieved 20% efficiency in silicon cells. The 2000 Nobel Prize in Physics was awarded in part to Zhores Alferov for semiconductor heterostructures, which enabled high-efficiency multi-junction solar cells. The 21st century saw rapid advances in thin-film, organic, and perovskite solar cells, reaching over 25% efficiency by 2020. As of end-2024, global installed PV capacity surpassed 2.2 terawatts (TW), with over 600 gigawatts (GW) added in 2024 alone, led by China, the EU, and the United States. Solar PV is now the fastest-growing and most cost-effective energy source worldwide, with commercial silicon modules exceeding 22% efficiency and tandem cells in labs reaching over 47%. With ongoing innovations in quantum materials, storage, and grid integration, solar energy is poised to dominate the global energy mix in the decades ahead.

1.7 Current Solar Cell Technologies

Solar cell technologies today span a wide range of materials and architectures, each tailored for specific performance, cost, and application needs. The dominant technology remains crystalline silicon (c-Si), which accounts for over 90% of the global market. These include monocrystalline and polycrystalline cells, with commercial efficiencies between 20 and 23%. Silicon's abundance, stability, and mature manufacturing ecosystem make it the industry standard. Thin-film technologies such as cadmium telluride (CdTe) and copper indium gallium selenide (CIGS) offer lightweight, flexible alternatives, especially for utility-scale and building-integrated PVs (BIPV).

A major leap in recent years has come from perovskite solar cells, which have achieved lab efficiencies exceeding 25% in under a decade. These are now being integrated into tandem configurations with silicon, pushing combined efficiencies beyond 30%. Multi-junction cells, used in space and concentrator PVs, have reached record efficiencies of over 47% in laboratory settings. These stack materials like gallium arsenide (GaAs) and indium gallium phosphide (InGaP) absorb different parts of the solar spectrum.

The National Laboratory of the Rockies (NLR) maintains the authoritative Best Research-Cell Efficiency Chart, which tracks the highest confirmed efficiencies across all PV technologies since 1976. As of 2026, the chart includes over 28 categories, from c-Si and thin films to emerging technologies like organic PVs (OPVs), quantum dots, and dye-sensitized solar cells (DSSCs). The chart highlights that tandem perovskite–silicon cells have surpassed 33%, while multi-junction III–V cells under concentrated sunlight exceed 47% ("Best Research-Cell Efficiency Chart" 1) [22] (Figure 1.8).

Emerging technologies are also gaining traction. Organic solar cells offer flexibility and low-cost processing, with efficiencies reaching 19%. Quantum dots and nanostructured solar cells are being explored for their tunable band gaps and potential in low-light environments. Meanwhile, bifacial modules, passivated emitter rear contact (PERC) cells, and heterojunction (HJT) technologies are enhancing silicon's performance in commercial products.

Today, solar PV is the fastest-growing energy source globally, with over 2.2 TW installed by the end of 2024 [23]. The field is advancing rapidly through innovations in materials science, nanotechnology, and device engineering, making solar energy more efficient, affordable, and versatile than ever before.

1.8 Chapter Summary

1) This chapter introduced the fundamental concept of the greenhouse effect, explaining how incoming solar radiation interacts with Earth's atmosphere and how greenhouse gases trap IR radiation, leading to global warming.
2) The major sources of global greenhouse gas emissions were discussed, with emphasis on the dominant role of the energy sector, followed by agriculture, industry, waste management, and land-use changes, highlighting the urgency of transitioning away from fossil fuels.
3) The observed impacts of climate change were examined through real-world examples, including rising global temperatures, sea-level rise, extreme weather events, ecosystem degradation, and region-specific case studies, demonstrating that climate change is already affecting both natural and human systems.

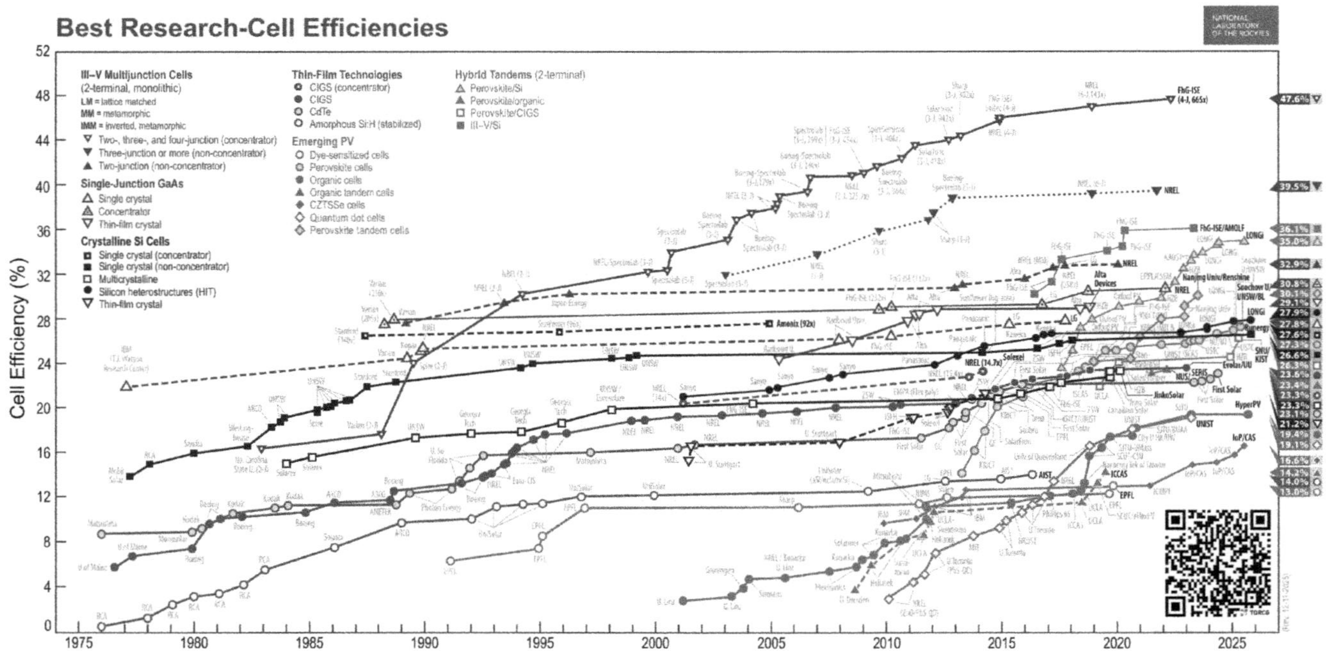

Figure 1.8 Best research solar cell efficiencies (1975–2025). *Source:* [22] / U.S. Department of Energy / Public domain. Scan QR code to access most recent solar cell efficiency table.

4) Solar energy was introduced as a renewable and sustainable energy source, with an explanation of the PV effect and the physical principles that enable the direct conversion of sunlight into electricity without greenhouse gas emissions during operation.
5) The economic viability of solar energy was analyzed by comparing installation costs and levelized cost of electricity across renewable technologies, showing that solar PV has become one of the most cost-effective sources of new electricity generation globally.
6) The role of solar energy in supporting energy independence and meeting rapidly increasing global energy demand was discussed, emphasizing its scalability, accessibility, and suitability for both developed and developing regions.
7) A brief historical overview traced the development of solar energy from early scientific discoveries to modern high-efficiency PV technologies, illustrating the rapid progress enabled by advances in physics, materials science, and semiconductor engineering.
8) Current solar cell technologies were reviewed, including c-Si, thin-film, perovskite, tandem, and multi-junction cells, highlighting recent efficiency records and the continued innovation driving the global expansion of solar PVs.

End-of-Chapter Problems

Conceptual Problems

P1.1 Explain how the greenhouse effect works and why an increase in greenhouse gas concentration leads to global warming rather than simply more reflected sunlight escaping into space.

P1.2 Why does the energy sector dominate global greenhouse gas emissions compared to agriculture, industry, and waste management, and what does this imply for climate mitigation strategies?

P1.3 Solar energy is described as abundant, scalable, and increasingly affordable. Which of these factors do you think has been most critical in accelerating global solar deployment, and why?

P1.4 Based on the historical development and current state of PV technologies discussed in this chapter, what factors do you think will most strongly influence the future growth of solar energy worldwide?

True/False Problems

P1.5 Solar energy conversion is possible because sunlight delivers energy in the form of photons.
True/False

P1.6 Greenhouse warming is mainly caused by Earth reflecting sunlight back into space.
True/False

P1.7 After absorbing sunlight, Earth re-emits part of that energy as infrared (IR) radiation.
True/False

P1.8 Greenhouse gases reduce how efficiently IR radiation escapes from Earth's atmosphere.
True/False

P1.9 The discovery of the photovoltaic effect predates modern silicon solar cells.
True/False

P1.10 In a solar cell, electrical power generation requires both light absorption and charge carrier collection.
True/False

P1.11 The AM1.5G spectrum represents sunlight outside Earth's atmosphere.
True/False

P1.12 A solar cell must be biased externally in order to generate photocurrent.
True/False

P1.13 The global shift toward clean energy is one reason solar photovoltaics has expanded rapidly.
True/False

References

1 Climate Watch, "Climate Watch (2025), with major processing by our world in data, 'Greenhouse gas emissions by sector' [dataset]." 2025. [Online].

Available: https://www.climatewatchdata.org/data-explorer/historical-emissions. [Accessed 2025].

2 X. Lan, P. Tans, K. Thoning, "Trends in globally-averaged CO2 determined from NOAA global monitoring laboratory measurements." 2025. [Online]. Available: https://doi.org/10.15138/9N0H-ZH07. [Accessed 2025].

3 NOAA Global Monitoring Laboratory, "Trends in atmospheric carbon dioxide (2025); EPA based on various sources (2022) – with major processing by our world in data." 2025. [Online]. Available: https://archive.ourworldindata.org/20251010-074813/grapher/co2-long-term-concentration.html. [Accessed 2025].

4 Climate Watch, "Climate watch, historical greenhouse gas (GHG) emissions [dataset]. Washington, D.C.: World Resources Institute (WRI)," 2025. [Online]. Available: https://www.climatewatchdata.org/data-explorer/historical-emissions. [Accessed 2025].

5 Intergovernmental Panel on Climate Change (IPCC), H. Lee, J. Romero, "Climate change 2023: synthesis report. Contribution of working groups I, II and III to the sixth assessment report of the intergovernmental panel on climate change. Geneva: IPCC," 2023. https://www.ipcc.ch/report/ar6/syr/downloads/report/IPCC_AR6_SYR_LongerReport.pdf. [Accessed 2025].

6 National Oceanic and Atmospheric Administration, "How does climate change affect coral reefs?" NOAA's National Ocean Service. https://oceanservice.noaa.gov/facts/coralreef-climate.html. [Accessed 2025].

7 K. Hall, *Bleaching of Coral Reefs Due to Ocean Acidification and Warming:Undergraduate Thesis*, Seattle, WA: University of Washington, 2021. https://digital.lib.washington.edu/researchworks/items/3ea3f9f3-3621-4778-bd8e-0214a529ce6a/full.

8 National Centre of Meteorology (UAE), "UAE witnesses largest rainfall in 75 years. Abu Dhabi: UAE Government News Agency (WAM).," 2024. https://www.wam.ae/en/article/13vbuq9-uae-witnesses-largest-rainfall-over-past-years. [Accessed 2025].

9 World Weather Attribution (WWA), "Heavy precipitation hitting vulnerable communities in the UAE and Oman becoming an increasing threat as the climate warms." 2024. [Online]. Available: https://www.worldweatherattribution.org/heavy-precipitation-hitting-vulnerable-communities-in-the-uae-and-oman-becoming-an-increasing-threat-as-the-climate-warms/. [Accessed 2025].

10 International Renewable Energy Agency, "Renewable power generation costs in 2023: executive summary. International Renewable Energy Agency (IRENA)," 2024. [Online]. Available: https://www.irena.org/-/media/Files/IRENA/Agency/Publication/2024/Sep/IRENA_Renewable_power_generation_costs_in_2023_executive_summary.pdf. [Accessed 2025].

11 IRENA, G. Nemet, J. D. Farmer et al., "Solar photovoltaic module price" part of the following publication: Hannah Ritchie, Pablo Rosado, and Max Roser

(2023) – 'Energy'." 2025. [Online]. Available: https://archive.ourworldindata.org/20250909-093708/grapher/solar-pv-prices.html. [Accessed 2025].

12 International Energy Agency, "Global energy review 2025. IEA," 2025. [Online]. Available: https://iea.org/reports/global-energy-review-2025. [Accessed 2026].

13 International Energy Agency, "Electricity – energy system (Data & statistics). Paris: IEA.," 2024. [Online]. Available: https://www.iea.org/energy-system/electricity. [Accessed 2025].

14 International Energy Agency, "Renewables 2023 – executive summary. Paris: IEA," 2024. [Online]. Available: https://www.iea.org/reports/renewables-2023/executive-summary. [Accessed 2025].

15 Our World in Data, "Ember (2025) – with major processing by Our World in Data. 'Electricity generation from other renewables, excluding bioenergy – Ember' [dataset]." 2025. [Online]. Available: https://ourworldindata.org/energy. [Accessed 2025].

16 Lafitte, A., Sordello, R., Ouédraogo, D.Y. et al. (2023). Existing evidence on the effects of photovoltaic panels on biodiversity: a systematic map with critical appraisal of study validity. *Environmental Evidence* 12: 25.

17 International Energy Agency (IEA), "World energy outlook (WEO) 2025." 2025. [Online]. Available: https://www.iea.org/reports/world-energy-outlook-2025?utm_content=bufferff04a&utm_medium=social&utm_source=linkedin-Birol&utm_campaign=buffer. [Accessed 2025].

18 Becquerel, A.E. (1839). Recherches sur les effets de la radiation chimique de la lumière solaire. *Comptes Rendus de L'Academie des Sciences* 9: 145–149.

19 Perlin, J. (2013). *Let It Shine: The 6000-Year Story of Solar Energy*. Novato, CA: New World Library.

20 Einstein, A. (1905). On a heuristic point of view about the creation and conversion of light. *Annalen der Physik* 17: 132–148.

21 The Nobel Prize Organization, "The Nobel prize in physics 1921 – Albert Einstein." 1921. [Online]. Available: https://www.nobelprize.org/prizes/physics/1921/einstein/facts/. [Accessed 2025].

22 National Laboratory of Rockies, "Interactive best research – cell efficiency chart." 2025. [Online]. Available: https://www.nlr.gov/pv/cell-efficiency. [Accessed 2025].

23 International Energy Agency, "Electricity – global energy review 2025 – analysis, Paris: IEA," 2025. [Online]. Available: https://www.iea.org/reports/global-energy-review-2025/electricity. [Accessed 2025].

2

Physics of Light

2.1 About the Chapter

In this chapter, we introduce the fundamental physics of light that is relevant to photovoltaic (PV) devices. We start by discussing the electromagnetic nature of light and the solar radiation spectrum, explaining the relationship between wavelength, frequency, and photon energy. We then discuss how incident photons interact with semiconductor materials, introducing the concepts of quantum efficiency (QE) and photogenerated current. Further, the propagation of light in materials is described using optical constants, including refractive index, n, and extinction coefficient, κ, which govern reflection and attenuation of the light when interacting with different materials. Building on this foundation and focusing on solar cell applications, we discuss thin-film interference and anti-reflective coatings (ARCs) and explain how optical losses at interfaces can be reduced. Finally, light propagation in multilayer structures is analyzed using the transfer matrix method (TMM), with and without considering the absorption. This provides students with a set of skills needed to calculate reflection, transmission, and absorption and to understand the link between optical design and solar cell device performance. By the end of this chapter, students are expected to gain the necessary tools needed to analyze and design light-management strategies in solar cells.

Key Topics

- Solar spectrum and photon energy
- Quantum efficiency and photocurrent J_{sc}
- Complex refractive index (n_s, κ) and absorption (α)
- Reflection and interference in thin films
- Quarter-wavelength
- Transfer matrix method (TMM) for multilayer optics

2.2 The Solar Radiation

Light propagates in the form of electromagnetic waves at a constant speed $c_0 = 3 \times 10^8$ m/s in free space. The range of optical wavelengths has three main regions: infrared (~0.8 to 300 μm), visible (~400 to 700 nm), and ultraviolet (~10 to 400 nm). Beyond these wavelengths, there are regions with microwave and radio waves (long wavelengths, with corresponding low frequency and low energy) to X-rays and gamma rays (short wavelengths, high frequency, and high energy). Figure 2.1 illustrates the electromagnetic spectrum and identifies the major spectral regions. At the short-wavelength, high-energy end of the spectrum is ionizing radiation. While ionizing radiation is abundant in outer space, the Earth's atmosphere absorbs most of it, thereby protecting life on the surface from its harmful effects.

As shown in Figure 2.1, electromagnetic radiation spans a wide range of wavelengths, extending from long-wavelength radio waves to high-energy gamma rays. In PV applications, we are mainly interested in the optical portion of the spectrum, which includes the ultraviolet, visible, and infrared regions, spanning between 10 nm and roughly 2–3 μm. These wavelengths correspond to photon energies comparable to electronic transitions in atoms and semiconductors, making them relevant for light–material interaction in optical devices.

Relationship between the wavelength λ, frequency ν, and photon energy $h\nu$ of electromagnetic wave is described by

$$h\nu = \frac{hc}{\lambda} \tag{2.1}$$

where h is Planck's constant, and c is the speed of light in vacuum. As highlighted by this relationship, we observe that shorter wavelengths correspond to higher photon energies. In semiconductor devices, only photons with energies exceeding the material bandgap, E_g, can be absorbed and generate electron–hole pairs, while lower-energy photons are transmitted or dissipated as heat. More details about this will be discussed in the following chapters of this book.

As solar radiation travels through Earth's atmosphere, its intensity decreases due to absorption by ozone, water vapor, and scattering caused by dust particles and aerosols in the air [1]. For instance, solar irradiance measures approximately 1353 W/m^2 in space, just outside the Earth's atmosphere, while this value drops to around 1000 W/m^2 when it reaches the Earth's surface. The impact of atmospheric conditions on the solar spectrum is characterized using the *air mass* (AM) *factor*, which quantifies the length of the optical path compared to the shortest possible path when the sun is positioned directly overhead. In simpler terms, the air mass corresponds to the secant of the angle formed between the sun's location and the zenith point. The AM 1 standard represents the spectrum when the

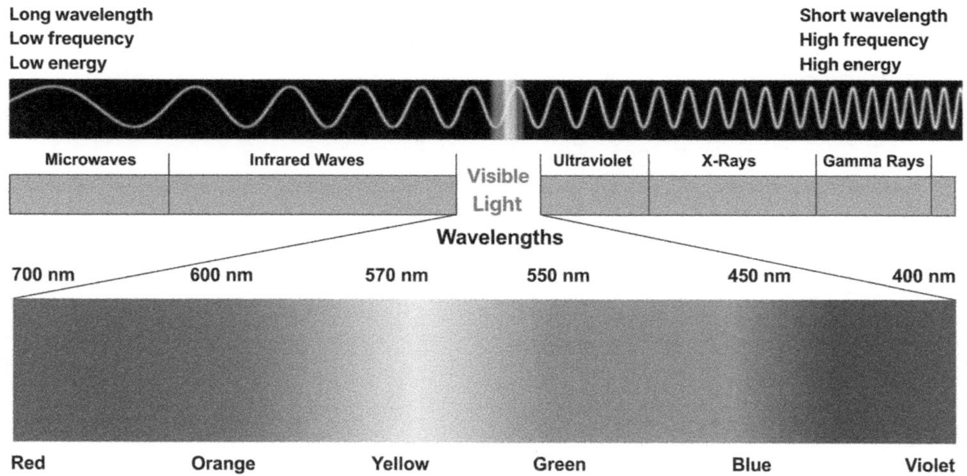

Figure 2.1 Electromagnetic wave spectrum with the visible light band highlighted.

sun is at zenith position (i.e. sun is right overhead, $\theta = 0°$), AM 1.5 standard corresponds to $\theta \approx 48.2°$, while AM 0 standard refers to the solar spectrum beyond the atmosphere.

The AM 1.5 standard is commonly specified in two forms: AM1.5G (global) and AM 1.5D (direct plus circumsolar). The AM 1.5G spectrum is used as the reference for typical outdoor solar irradiation under one-sun conditions and includes both the direct solar beam and the diffuse radiation scattered by the atmosphere. On the other hand, the AM 1.5D spectrum represents only the direct and circumsolar components of sunlight and is therefore used as the standard for concentrating solar power and concentrated PV (CPV) systems. Optical radiation incident on a surface is wavelength-dependent and commonly described in terms of its spectral distribution. The AM 0, AM 1.5G, and AM 1.5D standard spectral irradiance, $S_\lambda(\lambda)$, are shown in Figure 2.2. The spectral irradiance, $S_\lambda(\lambda)$, describes the incident optical power per unit area per unit wavelength, expressed in units of W m^{-2} nm^{-1}. The total incident power density is obtained by integrating $S_\lambda(\lambda)$ over wavelength. For the AM 1.5G spectrum, the total incident power density is approximately 1000 W/m^2 and about 900 W/m^2 for the AM 1.5D spectrum.

In certain situations, especially when calculating current generated by incident light, it is convenient to express solar irradiance in terms of the *photon flux density*, $\Phi(\lambda)$, defined as the number of incident photons per unit area per unit

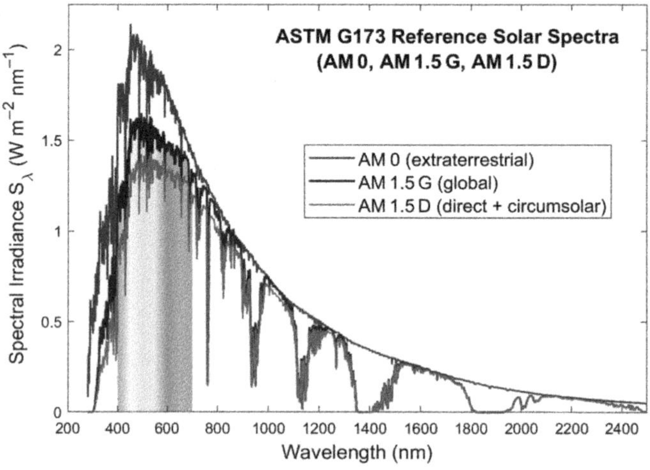

Figure 2.2 Reference solar spectral irradiance for extraterrestrial conditions (AM 0), standard global terrestrial conditions (AM 1.5G), and direct plus circumsolar terrestrial conditions (AM 1.5D) plotted as a function of wavelength. The shaded region indicates the visible spectral range (400–700 nm). Data are based on the (ASTM) G173 standard reference spectra, generated using the SMARTS radiative transfer model and distributed by the National Laboratory of the Rockies (NLR). *Source:* Adapted from [2].

time per unit wavelength (*#photons* $\cdot$ m^{-2} s^{-1} nm^{-1}). The relationship between spectral irradiance and photon flux density is given by

$$\Phi(\lambda) = S_\lambda(\lambda)\frac{\lambda}{hc} \tag{2.2}$$

This relationship becomes critical when evaluating photocurrent generation in solar cells, which depends on photon number rather than optical power alone.

From a practical perspective and applications to solar cells, the solar spectrum places an upper limit on how much current a solar cell can deliver, even before semiconductor material properties or any electrical losses are considered. This is because only photons with energies greater than the semiconductor bandgap, E_g, can generate electron–hole pairs that contribute to photogenerated current, while photons with lower energy pass through the device without contributing to current generation. This loss is unavoidable, and it depends on the material bandgap. Furthermore, photons with energies well above the bandgap, after being absorbed, have their excess energy lost through thermalization. Moreover, the high-energy photons, such as those in UV wavelength range, can cause structural damage to the semiconductor material, further reducing photogenerated current. These spectral losses are independent of device design and set the baseline efficiency limits that motivate optical optimization strategies in solar cells.

In the following sections of this chapter, we examine other factors that determine the amount of current that is generated by a solar cell. A convenient way to quantify how efficiently incident photons are converted into electrical current is through the QE, which is discussed in the next section.

2.3 Quantum Efficiency and Photogenerated Current

When light is absorbed in semiconductor material, photon energy generates free electron–hole pairs, which are then collected and contribute to photogenerated current. However, once light reaches a solar cell, not every incident photon ends up contributing to the current. Some photons are reflected at the surface before they ever enter the material, others travel through the device without being absorbed (they are transmitted), and some are absorbed but still fail to produce carriers that are collected at the contacts. There are many factors that play significant roles in how light is transformed into current, but they are often difficult to fully decouple from each other. So, to keep track of how many of the incident photons lead to useful electrical carriers, we introduce the concept of QE.

At its core, QE answers a very simple question: *out of all the photons arriving at the device, how many generate carriers that contribute to photogenerated current?*

Since both the light source and the material response depend strongly on wavelength of incident light, QE is treated as a wavelength-dependent quantity.

2.3.1 Quantum Efficiency

The **external QE (EQE)** at a given wavelength λ is defined as:

$$\mathrm{EQE}(\lambda) = \frac{\text{number of collected charge carriers per second}}{\text{number of incident photons per second}} \tag{2.3}$$

EQE includes all optical and electrical losses, such as surface reflection, incomplete absorption, and carrier recombination inside the device. An EQE value of 1 (or 100%) at a given wavelength means that each incident photon generates one electron–hole pair that is successfully collected, resulting in electric current. However, in practice, EQE is always less than unity due to the various loss mechanisms present in real devices.

Another useful measure is the internal QE (IQE), which is defined as the ratio of collected charge carriers to absorbed photons, rather than to incident photons. In this definition, optical losses due to surface reflection are excluded. When transmission through the device is negligible (i.e. most photons entering device are absorbed), the relationship between IQE and EQE can be approximated as

$$\mathrm{EQE} \approx (1 - R)\mathrm{IQE} \tag{2.4}$$

where R is wavelength-dependent reflectance.

While IQE provides insight into carrier transport and recombination, EQE is more directly measurable and is the quantity most often used when calculating device current under illumination. For this reason, the discussion in this chapter focuses primarily on EQE.

For solar cell to generate current, certain sequence of physical processes must occur:

- The photon must enter the device rather than being reflected at the surface.
- The photon must be absorbed within an active region of the material.
- The absorbed photon must generate an electron–hole pair.
- The generated carriers must reach the contacts to be collected.

Each of these steps depends on wavelength and has different behaviors, depending on the solar cell material and design. For example, short-wavelength photons are typically absorbed very close to the surface, where surface recombination may be significant and cause loss of generated carriers. On the other hand, long-wavelength photons may pass through the device without being absorbed, either due to their low energy or because material does not absorb well in this range. In addition, the solar cell can simply be too thin to absorb it. EQE is affected by all these factors, and as a result, the spectra often shows a strong variation across the optical spectrum (i.e. at different wavelengths).

From a practical perspective, an EQE curve provides immediate insight into where the optical or electrical losses are occurring. For example, a sharp drop in EQE at short wavelengths may indicate surface losses, while reduced EQE at long wavelengths often points to insufficient absorption or limited device thickness. These types of graphs help scientists and engineers decide which parts of solar cell design require improvement.

Figure 2.3 shows an example of the EQE spectrum of a silicon solar cell for different values of surface recombination velocity. The blue curve (triangles) corresponds to the highest surface recombination, which is why it has the lowest EQE at shorter wavelengths. In this wavelength range, carriers are generated very close to the front surface, so many of them recombine before they can be collected.

At longer wavelengths, absorption happens deeper inside the solar cell absorber, so surface recombination does not play such a strong role anymore. That is why all cases begin to overlap and show nearly the same EQE values in the long-wavelength region.

One interesting point in this example is that even in the ideal case with no surface recombination losses ($S_f = 0$ cm/s), the EQE is still below 1.0 at long wavelengths (roughly 700–1000 nm). This tells us that some of the incident light is still not being converted efficiently into collected photogenerated current. One obvious culprit here is the reflectance spectrum, *R*, which steadily increases as wavelength increases, meaning that more light is reflected away before it can even enter the device.

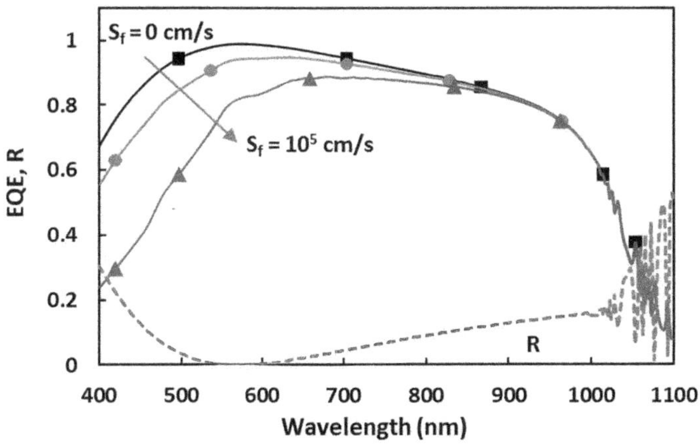

Figure 2.3 EQE characteristics of a silicon solar cell for different values of surface recombination velocity (reproduced with permission from [3]). The reflectance spectrum, *R*, for the same solar cell is also shown.

2.3.2 Photon Flux and Photogenerated Current

To relate QE to photogenerated electrical current, we describe the incident light in terms of photon flux rather than optical power. As introduced in Section 2.1, the photon flux density $\Phi(\lambda)$ represents the number of photons incident per unit area, per unit time, per unit wavelength (Eq. (2.2)).

Because each absorbed photon can generate at most one electron–hole pair, the contribution of photons at wavelength λ to the current density depends on both the photon flux and the QE of a solar cell at that wavelength.

The total photogenerated current density is obtained by summing the contributions from all wavelengths in the relevant spectral range and is calculated using the following relationship:

$$J_{ph} = q \int_{\lambda_{\min}}^{\lambda_{\max}} \Phi(\lambda)\, \mathrm{EQE}(\lambda)\, d\lambda \tag{2.5}$$

where q is the charge of the electron.

This equation tells us that the photocurrent is determined by the combination of the incident spectrum and the device response. Even if a device has excellent carrier transport and good electrical quality, it will still produce a small current if the QE is low over a large part of the spectrum. On the other hand, strong absorption alone is insufficient because carriers can still recombine before they are collected at the contacts.

In practical calculations, the integration limits are chosen to cover the spectral range where the photon flux is non-negligible, and the device exhibits a nonzero QE, which is also aligned with the bandgap of the absorbing material. Recall that the material absorbs only photons with energy greater than the absorbing material bandgap $h\nu \geq E_g$. For solar cells, this typically spans from the ultraviolet through the near-infrared, depending on the material and its semiconductor bandgap.

QE also provides a clear way to visualize fundamental spectral losses. Photons with energies below the bandgap cannot generate electron–hole pairs. This results in a sharp cutoff in the QE spectrum at long wavelengths for sufficiently thick absorber layer, corresponding to the bandgap energy of the material. On the other hand, as we have seen in the earlier example, QE can be low at short wavelengths, due to surface reflection or recombination effects, even though photon energies are large and well above the bandgap. These losses arise from device design and material quality, thus requiring optimization either in light absorption or electric transport. By examining how QE varies with wavelength, we can directly connect material properties, device structure, and optical design to the resulting current output. Optical engineering techniques such as antireflection coatings and surface passivation play a critical role in improving QE.

For this reason, the remainder of this chapter focuses on the optical physics underlying QE, such as reflection, absorption, and transmission, that we have mentioned. In the following sections, we start discussion with wave propagation in homogeneous media and progress to multilayer interference and absorption. We develop the tools needed to calculate reflection, transmission, and absorption spectra under different conditions.

Example 2.1 ***Current density from photon flux and EQE***
A silicon solar cell is illuminated by a monochromatic LED source that emits light at four discrete wavelengths. The incident photon flux density per unit wavelength, Φ, and the cell EQE at those wavelengths are given in the following table. Assume each wavelength band is narrow enough that Φ and EQE are constant across the band. The width of each band is also given in the table.

Band	Wavelength (nm)	Bandwidth $\Delta\lambda$ (nm)	Φ (photons $m^{-2}\ s^{-1}\ nm^{-1}$)	EQE
1	450	20	2.0×10^{20}	0.70
2	550	30	3.0×10^{20}	0.85
3	650	30	2.5×10^{20}	0.90
4	850	40	1.2×10^{20}	0.60

a) Calculate the short-circuit current density J_{sc} in A/m^2.
b) Convert your answer to mA/cm^2.
c) Which wavelength band contributes the most to J_{sc}, and why?

Solution
To calculate current density, we will use a discrete equivalent of Eq. (2.5)

$$J_{sc} = q \sum_i \Phi(\lambda_i)\ \mathrm{EQE}(\lambda_i)\ \Delta\lambda_i$$

with $q = 1.602 \times 10^{-19}$ C.

We can compute each band's contribution separately:

$$J_i = q\ \Phi_i\ \mathrm{EQE}_i\ \Delta\lambda_i$$

$$J_1 = (1.602 \times 10^{-19})(2.0 \times 10^{20})(20)(0.70) = 4.4856 \times 10^2\ \mathrm{A/m^2}$$

$$J_2 = (1.602 \times 10^{-19})(3.0 \times 10^{20})(30)(0.85) = 1.22553 \times 10^3\ \mathrm{A/m^2}$$

$$J_3 = (1.602 \times 10^{-19})(2.5 \times 10^{20})(30)(0.90) = 1.08135 \times 10^3\ \mathrm{A/m^2}$$

$$J_4 = (1.602 \times 10^{-19})(1.2 \times 10^{20})(40)(0.60) = 4.61376 \times 10^2\ \mathrm{A/m^2}$$

a) Total

$$J_{sc} = J_1 + J_2 + J_3 + J_4$$

$$J_{sc} = 448.56 + 1225.53 + 1081.35 + 461.38 = 3216.82\ \mathrm{A/m^2}$$

b) Convert to mA/cm^2
Use $1\ A/m^2 = 10^{-4}\ A/cm^2 = 0.1\ mA/cm^2$, so

$$J_{sc} = 3216.82 \times 0.1 = 321.68\ mA/cm^2$$

c) Largest contribution is from Band 2 because it has the largest product

$$\Phi \Delta\lambda \times EQE$$

2.4 Optical Properties of Materials

Before we proceed with a discussion of reflection and transmission, it is necessary to introduce the optical parameters that describe how light behaves inside a semiconductor. When light enters a material, the wave propagates through at a different speed compared to free space. Moreover, a part of the wave energy may be absorbed as it travels through a semiconductor. Both effects are captured by the *complex refractive index*, written as:

$$\tilde{n}(\lambda) = n_s(\lambda) + j\kappa(\lambda) \tag{2.6}$$

The real part of Eq. (2.6), n_s, is the *refractive index* that controls the phase velocity of light inside the material. It is often defined as the ratio between the speed of light in vacuum, c_o, and the speed of light inside the medium, c

$$n_s = c_o/c \tag{2.7}$$

The imaginary part of Eq. (2.6), κ, is called the *extinction coefficient*, and it accounts for optical absorption. Absorption is often described using the *absorption coefficient*, α, which depends directly on κ. The relationship is

$$\alpha(\lambda) = \frac{4\pi\kappa(\lambda)}{\lambda} \tag{2.8}$$

where λ is the wavelength of the incident light.

Figure 2.4 shows refractive indices, n_s, and extinction coefficients for some common materials used in solar cells. Figure 2.5 shows absorption coefficients for the same group of materials. It is important to note that both n_s and κ depend on wavelength. Since the solar cell is illuminated by a wide spectrum of wavelengths, the reflection losses at interfaces and the absorption inside each layer change across the spectrum. Optical modeling, therefore, aims to compute the wavelength-dependent reflectance $R(\lambda)$, transmittance $\mathcal{T}(\lambda)$, and absorption $A(\lambda)$.

A useful way to understand the absorption coefficient is through the absorption depth (also called penetration depth), defined as

$$\delta = \frac{1}{\alpha} \tag{2.9}$$

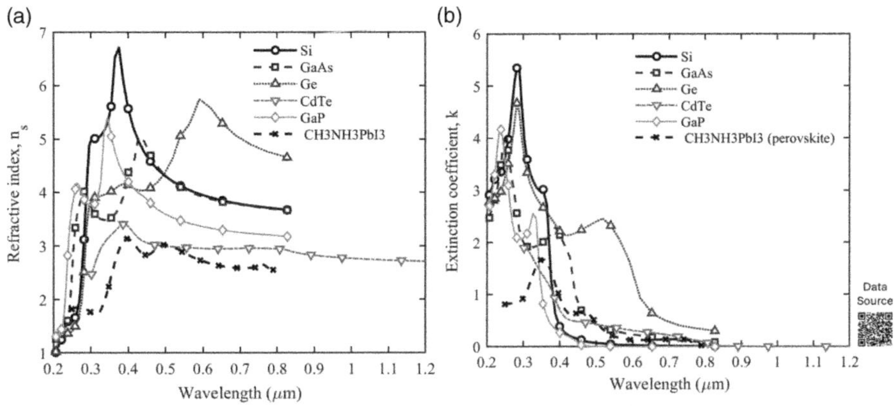

Figure 2.4 Optical properties of materials used in optical devices: (a) refractive index n_s and (b) extinction coefficient, κ, as a function of wavelength. *Source:* Adapted from [4].

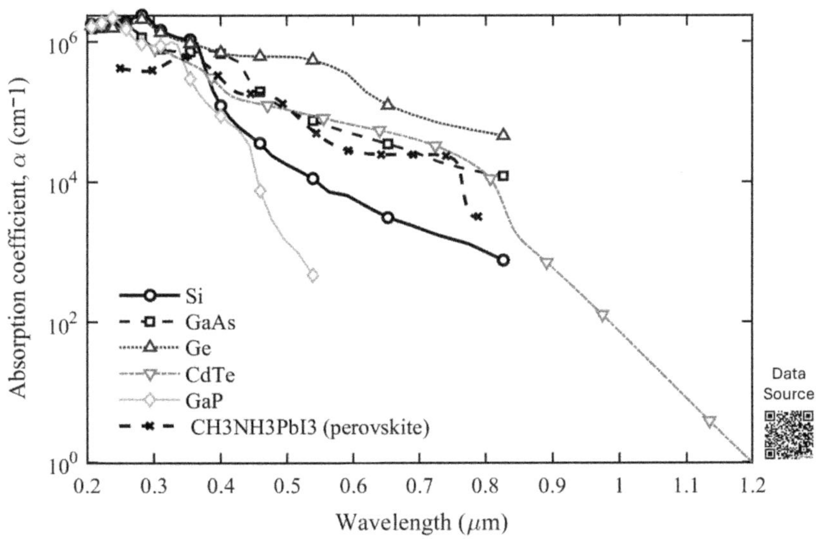

Figure 2.5 Absorption coefficients vs. wavelengths (μm) for materials often used in optical devices. *Source:* Adapted from [4].

Absorption depth represents the distance into the material over which the optical intensity drops to $\frac{1}{e}$ (about 37%) of its value at the surface, assuming reflection is neglected.

Once light enters the semiconductor, its intensity decreases with propagation depth because part of the optical energy is absorbed. If we ignore reflection at the interface ($R = 0$) for the time being, the attenuation of the light intensity through the depth, x, reduces exponentially, as described below:

$$\mathbb{I}(x) = \mathbb{I}_0 e^{-\alpha x} \tag{2.10}$$

where $\mathbb{I}_0$ is the incident intensity at the surface, and $\mathbb{I}(x)$is the light intensity at depth x.

Example 2.2 ***Absorption coefficient and absorption depth from* k**
A semiconductor has an extinction coefficient of $\kappa = 0.25$ at a wavelength of $\lambda = 600$ nm.

a) Calculate the absorption coefficient α in cm^{-1} using
b) Calculate the absorption depth δ in nm, defined as

Use: 1.0 nm $= 10^{-7}$ cm.

Solution
Given: $\kappa = 0.25$, $\lambda = 600$ nm $= 6 \times 10^{-5}$ cm

a) $$\alpha = \frac{4\pi\kappa}{\lambda} = \frac{4\pi(0.25)}{6 \times 10^{-5}} = \frac{\pi}{6 \times 10^{-5}} = 5.24 \times 10^4\ \text{cm}^{-1}$$

b) $$\delta = \frac{1}{\alpha} = \frac{1}{5.24 \times 10^4} = 1.91 \times 10^{-5}\ \text{cm}$$

Convert to nm:

$$1.91 \times 10^{-5}\ \text{cm} \times (10^7\ \text{nm/cm}) = 191\ \text{nm}$$

Equation (2.10) is known as Beer–Lambert law, and it tells us how light absorption behaves at a given depth for a specific absorption coefficient, α. For example, when α is large, absorption happens very close to the surface, while for small α, photons can penetrate deeper into the material before being absorbed. This matters in solar cells because the location of absorption determines where free carriers are generated, and this affects the likelihood of these carriers being collected before recombination occurs.

With the complex refractive index defined, we are now ready to discuss reflection and transmission at interfaces. As we will see, both n_s and κ influence how much light enters the device, and therefore how much of the incident spectrum remains available for useful absorption.

2.5 Reflectance, Transmittance, and Absorption

The reflectance, R, at a surface is defined as the ratio of the reflected optical power ($\mathbb{I}_r$) to the optical power incident on that surface ($\mathbb{I}_i$). Similarly, transmittance, ($\mathscr{T}$),

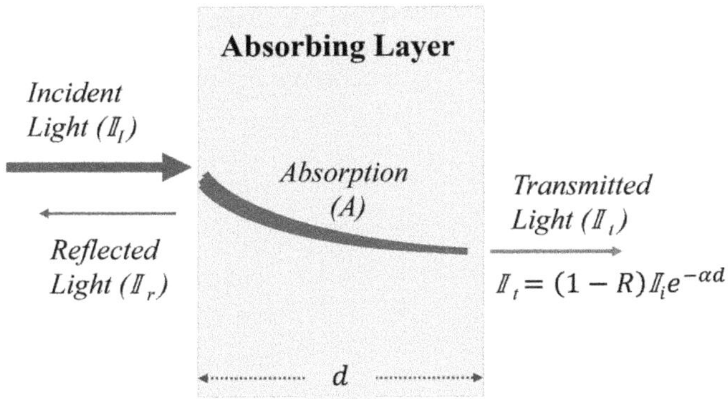

Figure 2.6 Illustration of the relationship between reflectance, *R*, absorption, *A*, and transmittance, $\mathcal{T}$.

is the ratio between transmitted power ($\mathbb{I}_t$) and incident power ($\mathbb{I}$). These relationships are defined as:

$$R = \frac{\mathbb{I}_r}{\mathbb{I}_i}, \quad \mathcal{T} = \frac{\mathbb{I}_t}{\mathbb{I}_i} \tag{2.11}$$

The light that is not reflected and that enters the layer is either transmitted or absorbed within it. The amount of absorbed light is defined as absorption, *A*. This leads us to the conservation of energy relationship,

$$R(\lambda) + A(\lambda) + \mathcal{T}(\lambda) = 1 \tag{2.12}$$

These relationships are illustrated in Figure 2.6, assuming normal incidence.

2.5.1 Reflectance and Transmittance at a Single Interface (Lossless, $A = 0$)

When light penetrates a structure that contains one or more interfaces between materials with different refractive indices, n_s, a portion of the incoming light is reflected at each interface. At a boundary between two materials with refractive indices n_1 and n_2, an incident ray of light generally splits into a reflected ray and a transmitted ray. The angles of the reflected and transmitted rays are related through Snell's law.

$$n_1 \sin\theta_1 = n_2 \sin\theta_2 \tag{2.13}$$

Figure 2.7 illustrates an interface between *Material* 1 with refractive index n_1 and *Material* 2 with refractive index n_2, with an incident light ray (wave) approaching the boundary. A wave traveling in *Material* 1 reaches the interface at $x = 0$ at an angle θ_1 relative to the normal to the interface. The reflected ray propagates back

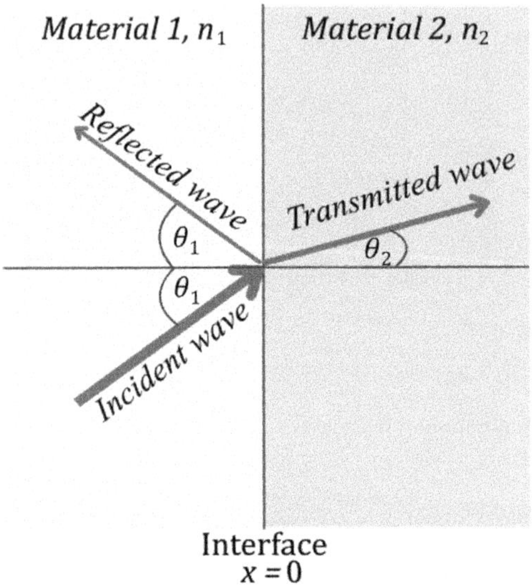

Figure 2.7 Reflection and transmission of an optical wave at an interface between two media (n_1, n_2).

into *Material* 1 at the same angle θ_1, while the transmitted ray propagates in *Material* 2 at an angle θ_2 with respect to the normal. The quantities n_1, n_2, θ_1, and θ_2 satisfy Snell's law (Eq. (2.13)).

The reflected and transmitted amplitudes are completely determined by the complex refractive indices of the two media (n_s and κ). For most optical design problems in PVs, reflection at the first interface is a major loss mechanism because semiconductor materials usually have high refractive indices compared to air or glass ($n_{glass} \approx n_{air} = 1$). This large mismatch is the reason uncoated silicon reflects a substantial fraction of incident sunlight (~30% on average) and why solar cells require ARC.

2.5.1.1 Reflection at a Single Interface (Normal Incidence, No Absorption)

Let us consider a plane wave at a specific wavelength that enters *Material* 1 at normal incidence (i.e. $\theta_1 = 0$ in Figure 2.7), traveling from *Material* 1 with refractive index n_1 into *Material* 2 with refractive index n_2. As we discussed earlier, the portion of the incident wave is reflected back, while the rest is transmitted forward. Ignoring the absorption for the time being, the amplitude reflection coefficient that describes the amount of reflection is defined by the Fresnel relation:

$$r_{12} = \frac{n_1 - n_2}{n_1 + n_2} \tag{2.14}$$

The reflectance, R, which represents reflected optical power fraction, is defined in terms of reflection coefficients as:

$$R = |r_{12}|^2 \tag{2.15}$$

Note that Eqs. (2.14) and (2.15) provide reflection coefficient, r_{12}, and reflectance, R, for one single wavelength entering at normal incidence and assuming no absorption. Also note that when two materials have matched refractive indices such as $n_1 \approx n_2$, then $r_{12} \approx 0$, meaning that reflectance is negligible and almost all incident light enters the *Material* 2. However, if $n_2 \gg n_1$, reflection becomes significant. This is exactly the main challenge with PV devices and the reason why we design ARCs as integral part of solar cells. This design topic will be discussed in more detail in Chapter 6.

Example 2.3 ***Air–Silicon Interface Reflection: Coefficient and Reflectance Calculation***

Consider the air/silicon–air interface with $n_{air} \approx 1$, and silicon refractive index at specific wavelength to be $n_{Si} \approx 3.5$. Calculate reflection coefficient $r_{air\text{-}Si}$ and reflectance $R_{air\text{-}Si}$ for light propagation from air to Si.

Solution

$$r_{air-Si} = \frac{n_{air} - n_{Si}}{n_{air} + n_{Si}} = \frac{1 - 3.5}{1 + 3.5} = -0.5556$$

Then $R_{air \to Si} \approx \left(\frac{1-3.5}{1+3.5}\right)^2 \approx 0.31$

The result of example 2.3 shows that 31% of incident light at this specific wavelength is reflected and never enters Si, which is a significant loss of power.

This is one of the main motivations for using ARCs, which reduce optical coupling losses and increase the number of photons available for absorption and carrier generation.

2.5.1.2 Transmittance at a Single Interface (Normal Incidence, No Absorption)

For normal incidence in lossless dielectrics, the transmittance is also defined by Fresnel expression as:

$$t_{12} = \frac{2n_1}{n_1 + n_2} \tag{2.16}$$

For lossless media at normal incidence, the transmittance as ratio of transmitted power can be defined as:

$$\mathscr{T} = \frac{n_2}{n_1}|t_{12}|^2 \tag{2.17}$$

Substituting Eq. (2.16) into Eq. (2.17) leads to

$$\mathscr{T} = \frac{n_2}{n_1}|t|^2 = \frac{4n_1 n_2}{(n_1 + n_2)^2} \tag{2.18}$$

For lossless medium (no absorption, $A(\lambda) = 0$), energy conservation relation comes down to $R(\lambda) + \mathscr{T}(\lambda) = 1$, which is a good sanity check when estimating transmission and reflection in a specific layer.

2.5.2 Transfer Matrix Method for Lossless Multilayer Optics

So far, we have seen examples with one interface, such as air to silicon. However, in practice, optical devices such as solar cells have many different layers, thus many interfaces, where each layer has different optical properties. Reflected waves at these interfaces can interfere with each other, either constructively or destructively depending on wavelength and layer thickness.

When a light wave encounters more than one interface, the reflection cannot be described using a single Fresnel equation anymore (i.e. Eqs. (2.15) and (2.17)). Instead, the wave partially reflects and transmits at each boundary, while also accumulating phase as it propagates through each layer. Instead of tracking individual reflected and transmitted rays at each interface, we can look at overall accumulated reflected and transmitted rays at each layer.

Inside any uniform layer, the optical field can be described as the sum of a forward-traveling wave, U^+, and a backward-traveling wave, U^-. These two amplitudes fully describe what the wave is doing at any point inside a uniform layer, even though physically the backward wave may be the result of many internal reflections.

Consider a plane wave incident from the left onto a structure made of multiple thin layers, each with its own refractive index and thickness, as illustrated in Figure 2.8. As the wave enters layer 1 at interface 1, part of it is reflected and part propagates through Layer 1. At interface 2, between layer 1 and layer 2, part of the wave is transmitted into the next region (layer 2), while another part is reflected backward (toward interface 1). That reflected portion can either reflect again at interface 1 or transmit back toward the front surface. Very quickly, even a simple two-layer stack such as shown below produces a sequence of multiple reflected and transmitted waves traveling in both directions that is difficult to keep track of. It is more convenient to represent the field inside each layer using only two net wave components: a forward-traveling wave amplitude U^+ and a backward-traveling wave amplitude U^-. In case of 2-layer stack in Figure 2.8, U_1^+ and U_2^+ represent forward-collected waves in layers 1 and 2, respectively, while U_1^- and U_2^- represent backward-collected waves in layers 1 and 2, respectively. The thickness of each layer, d_i, controls how much phase the wave accumulates as it propagates, while the refractive index mismatch at each boundary controls how strongly the wave reflects. This physical picture, forward and backward waves linked across interfaces and modified by phase accumulation in each layer, forms the foundation for the TMM introduced next.

The TMM provides a compact way to analyze reflection and transmission in multilayer structures. The full stack response is obtained by multiplying matrices in the specific order that the incident wave encounters the layers. Considering two subsequent layers, their relationship can be represented by matrix equation:

$$\begin{bmatrix} U_1^+ \\ U_1^- \end{bmatrix} = M \begin{bmatrix} U_2^+ \\ U_2^- \end{bmatrix}, \quad M = \begin{bmatrix} A & B \\ C & D \end{bmatrix} \tag{2.19}$$

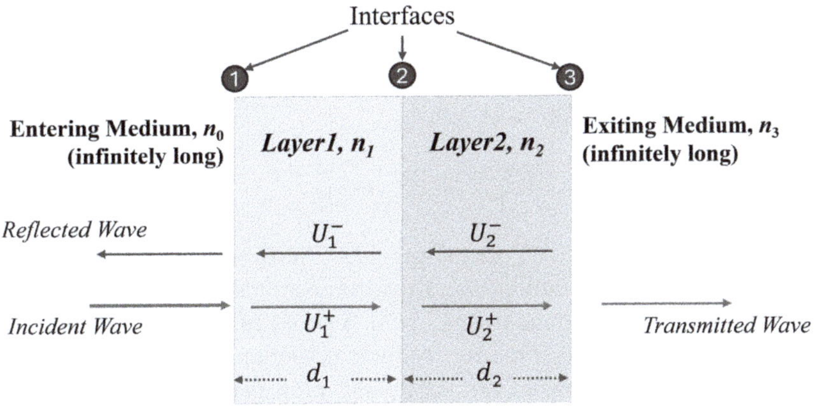

Figure 2.8 Forward and backward-traveling wave amplitudes in a multilayer structure. In each layer, the optical field is represented as the sum of a forward wave U^+ and a backward wave U^-. The effect of a multilayer stack is obtained by relating these amplitudes across the interfaces (points 1, 2, and 3) while accounting for phase accumulation through each thickness d_i.

The matrix M, with elements A, B, C, and D, is called wave transfer matrix.

Let us now consider a multilayer structure made of N layers, sandwiched between two semi-infinite media. We denote the incident medium as medium 0 with refractive index n_0, and the exit medium as medium $N+1$ with refractive index n_{N+1}. Each layer i has refractive index n_i and thickness d_i. As the wave propagates through a homogeneous layer i, it accumulates a phase delay, $\varphi_i = n_i d_i \frac{2\pi}{\lambda}$. In matrix form, the wave propagation through layer i is defined as:

$$M_{\text{prop},i} = \begin{bmatrix} e^{-j\varphi_i} & 0 \\ 0 & e^{j\varphi_i} \end{bmatrix}, \quad \varphi_i = n_i d_i \frac{2\pi}{\lambda} \tag{2.20}$$

This matrix accounts for phase accumulation only. For the case when absorption is ignored, as in this section, the wave amplitude does not decay inside the layer. The parameter φ_i is sometimes called the phase thickness of the layer, and it depends on wavelength, thickness, and refractive index.

We next analyze reflectance at each interface. At interface between two neighboring layers i and $(i+1)$, the wave experiences partial reflection and transmission due to the difference in refractive indices. This boundary can also be represented by a transfer matrix, written as:

$$M_{i/(i+1)} = \frac{1}{2n_{(i+1)}} \begin{bmatrix} n_{(i+1)} + n_i & n_{(i+1)} - n_i \\ n_{(i+1)} - n_i & n_{(i+1)} + n_i \end{bmatrix} \tag{2.21}$$

It should be noted that this interface matrix carries the same information as Fresnel reflection and transmission coefficients, but it is written in a form that becomes convenient once multiple layers are present.

A wave propagating through N layers will encounter a sequence of interfaces and propagation steps, described by $M_{i/(i+1)}$ and $M_{\text{prop},i}$, respectively. The total transfer matrix M of the multilayer stack is found by multiplying the matrices of all interfaces and layers in the correct order:

$$M = \begin{bmatrix} A & B \\ C & D \end{bmatrix} = \prod_{i=1}^{N} (M_{(i-1)/i} M_{\text{prop},i} M_{i/(i+1)}) \tag{2.22}$$

Combining Eq. (2.22) with Eq. (2.19) provides the information about total reflected wave in entering medium 0 and total transmitted wave in exiting the medium. This result is the main reason TMM is so useful in solar cell optics. Once the stack is described as a matrix product, adding another optical layer is straightforward. We simply insert one additional propagation matrix and two additional interface matrices. It is important to note that all elements of matrix M depend on wavelength λ of the incoming wave.

From Eq. (2.22), we can calculate the reflectance of the entire structure, seen by the incident wave in medium 0, and it can be written as

$$R_{0,(N+1)} = \left| \frac{-C}{D} \right|^2 \tag{2.23}$$

This expression gives the reflectance of the full stack as a function of wavelength. It automatically accounts for all internal reflections and interference effects caused by the multiple interfaces. For a lossless case, we are analyzing here ($A = 0$), transmittance, $\mathcal{T}$, through entire stack can be found by $\mathcal{T} = 1 - R_{0,(N+1)}$.

In the following examples, we will apply these equations to thin-film coatings used in PV devices. We will start with a single-layer ARC and show how the quarter-wavelength condition appears naturally through the phase term φ_i. Then, we will extend the method to multilayer coatings, where broadband reflection suppression can be achieved.

Example 2.4 ***Single interface (air/Si) using TMM***

Compute reflectance R for a single boundary air → silicon using the TMM and verify it matches the Fresnel result. Assume $n_1 = 3.5$ at a given wavelength.

Solution

Step 1: Define the Structure This is a "stack" with **zero layers** between two semi-infinite media:

- medium 0: air, $n_0 = 1$
- medium 1: silicon, $n_1 = 3.5$

There is only one interface 0/1, so the total matrix is simply the interface matrix:

$$M = M_{0/1}$$

Step 2: Write the Interface Matrix Using Eq. (2.21):

$$M_{0/1} = \frac{1}{2n_1}\begin{bmatrix} n_1 + n_0 & n_1 - n_0 \\ n_1 - n_0 & n_1 + n_0 \end{bmatrix}$$

Substitute $n_0 = 1$, $n_1 = 3.5$:

$$M = \frac{1}{2(3.5)}\begin{bmatrix} 3.5+1 & 3.5-1 \\ 3.5-1 & 3.5+1 \end{bmatrix} = \frac{1}{7}\begin{bmatrix} 4.5 & 2.5 \\ 2.5 & 4.5 \end{bmatrix}$$

So

$$A = \frac{4.5}{7}, \quad B = \frac{2.5}{7}, \quad C = \frac{2.5}{7}, \quad D = \frac{4.5}{7}$$

Step 3: Use the Reflectance Formula from TMM Using Eq. (2.23):

$$R_{0,1} = \left|\frac{-C}{D}\right|^2$$

$$R_{0,1} = \left|\frac{-\frac{2.5}{7}}{\frac{4.5}{7}}\right|^2 = \left|\frac{-2.5}{4.5}\right|^2 = \left(\frac{2.5}{4.5}\right)^2$$

$$R_{0,1} = \left(\frac{0.5556}{1}\right)^2 \approx 0.3086$$

Step 4: Compare to Fresnel Result *From the Fresnel equation at normal incidence:*

$$R = \left(\frac{n_0 - n_1}{n_0 + n_1}\right)^2 = \left(\frac{1-3.5}{1+3.5}\right)^2 = \left(\frac{-2.5}{4.5}\right)^2 \approx 0.31$$

So, the TMM result matches exactly, as expected.

Example 2.5 ***Quarter-wavelength ARC using TMM (Air/SiO_2/Si)***

Use TMM to show how a SiO_2 film on Si can reduce reflection at a chosen wavelength by choosing a quarter-wavelength thickness.

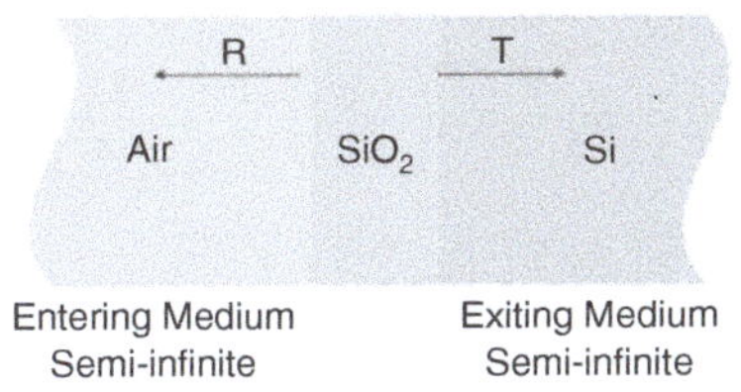

Assume the following

- the design wavelength: $\lambda_0 = 600$ nm
- medium 0: air, $n_0 = 1$
- layer 1: SiO_2, $n_1 = 1.46$, thickness d_1
- medium 2: silicon, $n_2 = 3.5$

Solution

We have one thin layer between air and silicon:

$$\text{air}\,(n_0)\ \mid\ SiO_2(n_1, d_1)\ \mid\ Si\,(n_2)$$

We need to find thickness of SiO_2, d_1, that minimizes reflection.

Step 1: Write the Total Transfer Matrix
Using Eq. (2.22), for one layer:

$$M = M_{0/1} \;\; M_{\text{prop},1} \;\; M_{1/2}$$

where

$$M_{\text{prop},1} = \begin{bmatrix} e^{-j\varphi_1} & 0 \\ 0 & e^{j\varphi_1} \end{bmatrix}, \quad \varphi_1 = n_1 d_1 \frac{2\pi}{\lambda}$$

and the two interface matrices come from Eq. (2.21).

Step 2: Choose Design Wavelength and Quarter-Wave Thickness
Let the design wavelength be

$$\lambda_0 = 600 \text{ nm}$$

A **quarter-wavelength coating** is defined by

$$d_1 = \frac{\lambda_0}{4n_1}$$

Substitute $n_1 = 1.46$:

$$d_1 = \frac{600}{4(1.46)} \approx \frac{600}{5.84} \approx 102.7 \text{ nm}$$

Quarter-wavelength thickness of SiO_2: $\boxed{d_1 \approx 103 \text{ nm}}$

Step 3: Evaluate the Phase Thickness at the Design Wavelength
At $\lambda = \lambda_0$:

$$\varphi_1 = n_1 d_1 \frac{2\pi}{\lambda_0}$$

Substitute $d_1 = \lambda_0/(4n_1)$:

$$\varphi_1 = n_1 \left(\frac{\lambda_0}{4n_1}\right) \frac{2\pi}{\lambda_0} = \frac{2\pi}{4} = \frac{\pi}{2}$$

So, for a quarter-wave layer:

$$\boxed{\varphi_1 = \frac{\pi}{2}}$$

This is the key point: the propagation matrix becomes

$$M_{\text{prop},1} = \begin{bmatrix} e^{-\frac{j\pi}{2}} & 0 \\ 0 & e^{\frac{j\pi}{2}} \end{bmatrix} = \begin{bmatrix} -j & 0 \\ 0 & +j \end{bmatrix}$$

That phase shift forces the two strongest reflected contributions (from the top and bottom of the ARC film) to come back with opposite phase, which is why reflection can drop sharply near λ_0.

Step 4: Compute Reflectance using the Matrix Result

Once

$$M = \begin{bmatrix} A & B \\ C & D \end{bmatrix}$$

is obtained from the multiplication

$$M = M_{0/1} \;\; M_{\text{prop},1} \;\; M_{1/2}$$

$$M_{0/1} = \frac{1}{2n_1}\begin{bmatrix} n_1 + n_0 & n_1 - n_0 \\ n_1 - n_0 & n_1 + n_0 \end{bmatrix}, \quad M_{1/2} = \frac{1}{2n_2}\begin{bmatrix} n_2 + n_1 & n_2 - n_1 \\ n_2 - n_1 & n_2 + n_1 \end{bmatrix}$$

Substituting values:

$$M = \frac{1}{2(1.46)}\begin{bmatrix} 1.46+1 & 1.46-1 \\ 1.46-1 & 1.46+1 \end{bmatrix}\begin{bmatrix} -j & 0 \\ 0 & +j \end{bmatrix}\frac{1}{2(3.5)}\begin{bmatrix} 3.5+1.46 & 3.5-1.46 \\ 3.5-1.46 & 3.5+1.46 \end{bmatrix}$$

$$M = \begin{bmatrix} -0.551j & -0.134j \\ 0.134j & 0.551j \end{bmatrix}$$

the reflectance is computed using Eq. (2.23):

$$R_{0,2} = \left|\frac{-C}{D}\right|^2 = \left|\frac{-0.134j}{0.551j}\right|^2 \approx 0.059$$

This gives the total reflectance of the full stack as seen from air.

Step 5: What Should We Expect Numerically?

Even without calculating the full matrix product by hand, we can predict the trend:

- bare Si reflects ~31% at normal incidence
- adding a quarter-wave SiO_2 layer near a visible wavelength reduces reflection significantly at λ_0

However, because SiO_2 does not satisfy the ideal index condition

$$n_1 = \sqrt{n_0 n_2} = \sqrt{1 \times 3.5} \approx 1.87$$

the reflection will not go all the way to zero. Still, it will drop strongly compared to the bare surface. So the quarter-wave thickness is a realistic and widely used ARC design rule, even though it is not always perfect.

Following up on Example 2.5, we note that destructive interference between the two reflected waves occurs when the round-trip phase shift equals π, which leads to

$$2\varphi = \pi \Rightarrow n_1 d = \frac{\lambda_0}{4} \tag{2.24}$$

This condition explains the name "quarter-wavelength coating": the physical thickness of the layer corresponds to one quarter of the wavelength inside the material, not in free space. At the design wavelength λ_0, the reflections from the two interfaces cancel, resulting in minimum reflectance.

2.5.3 Anti-Reflective Coatings and the Quarter-Wavelength Condition

ARCs are thin dielectric layers deposited on the surface of a device to suppress reflection through wave interference rather than absorption. Their operation relies on the phase relationship between multiple reflected waves, rather than modifying the material properties of the absorber itself.

In addition to satisfying the quarter-wavelength thickness, the amplitudes of the two reflected waves must be comparable. This is achieved when the refractive index of the coating is approximately the geometric mean of the refractive indices of the surrounding media,

$$n_1 \approx \sqrt{n_0 n_s} \tag{2.25}$$

For an air–silicon interface, this ideal value lies near $n_1 \approx 1.9$. In practice, common dielectric materials such as SiO_2 ($n \approx 1.46$) or SiN_x ($n \approx 1.9 - 2.1$) are used, depending on fabrication constraints and spectral requirements. When the coating index deviates from the ideal value, the reflectance minimum at the design wavelength is no longer zero, but can still be reduced. It should be noted that a single-layer ARC is wavelength-selective. In other words, reflection is minimized only near the design wavelength, while reflectance increases at shorter and longer wavelengths. Since the solar spectrum spans a wide wavelength range, a quarter-wavelength ARC represents a compromise rather than a broadband solution. Often, engineers will minimize reflection at a wavelength that contributes most to photogenerated current, which can be observed from QE spectra.

Despite this limitation, single-layer ARCs are widely used because they are simple and inexpensive, and their performance can be enhanced by surface texturing, which is discussed in later sections.

2.5.4 Limitations of Lossless TMM

While the lossless TMM captures interference and reflection accurately, it does not describe absorption within the layers. As a result, it cannot predict how much light is absorbed in the active regions of a device, nor can it determine photocurrent or QE. Nevertheless, this simplified model is very useful for optical design. By isolating phase and interference effects, it allows ARCs and optical stacks to be optimized without conflating optical losses with material quality or carrier transport effects.

2.6 Multilayer Propagation with Absorption

In the previous sections, we developed the optical tools needed to describe reflection and interference in layered structures. We started with reflection at a single interface, then extended the discussion to thin-film interference and quarter-wavelength layers as a practical way to reduce reflection. Up to this point, we treated the layers as lossless, meaning that light could propagate through each layer without being absorbed and attenuated. This assumption is useful for building intuition; however, in reality optoelectronic devices are not lossless. In solar cells, absorption is the main reason the device generates current, and it is significant.

In this section, we include absorption in a multilayer structure, so that our analysis can tell us how much light is absorbed, and more importantly, where that absorption happens inside the stack.

In this section, we extend the multilayer description to include absorption explicitly while keeping the same framework and notation used earlier.

As introduced in Section 2.3, absorption in optical materials is described through the *complex refractive index*

$$\tilde{n}(\lambda) = n_s(\lambda) + j\kappa(\lambda)$$

where n_s controls phase propagation and k is the extinction coefficient that accounts for absorption (i.e. attenuation) of the wave.

When light propagates through a layer of thickness d_i, the phase term introduced earlier becomes a function of complex refractive index, $\tilde{n}$, as:

$$\varphi_i = \tilde{n}_i d_i \frac{2\pi}{\lambda} \tag{2.26}$$

This looks similar to the lossless case, except that $\tilde{n}_i$ is now complex. Because of that, φ_i also becomes complex, meaning that the same propagation term now describes two effects at once: the wave accumulates phase as it travels forward, and its amplitude decays due to absorption.

2.6.1 Transfer Matrix Method with Complex Refractive Index

The TMM derived earlier remains valid when absorption is present. The structure is still represented as a sequence of interface matrices and propagation matrices. The only change is that each layer now uses the complex refractive index $\tilde{n}_i$ rather than a real index n_i.

The propagation matrix for layer i is defined using the same form as in the lossless case (Eq. (2.20)), but with phase defined as in Eq. (2.26), and is written as

$$M_{\text{prop},i} = \begin{bmatrix} e^{-j\tilde{n}_i d_i \frac{2\pi}{\lambda}} & 0 \\ 0 & e^{j\tilde{n}_i d_i \frac{2\pi}{\lambda}} \end{bmatrix} \tag{2.27}$$

The interface matrix between two neighboring layers i and $(i+1)$ keeps the same form as before, but with complex indices:

$$M_{i/(i+1)} = \frac{1}{2\tilde{n}_{(i+1)}} \begin{bmatrix} \tilde{n}_{(i+1)} + \tilde{n}_i & \tilde{n}_{(i+1)} - \tilde{n}_i \\ \tilde{n}_{(i+1)} - \tilde{n}_i & \tilde{n}_{(i+1)} + \tilde{n}_i \end{bmatrix} \tag{2.28}$$

For a stack of N layers between two semi-infinite media (n_0 and n_{N+1}), the total matrix is still written as Eq. (2.22).

The main point to emphasize here is that absorption does not "break" interference. The wave still reflects back and forth and accumulates phase, but the wave traveling deeper into the structure becomes progressively weaker. This means the contributions coming from deeper interfaces become less important once attenuation is strong. That is why thin-film interference effects are typically more pronounced in transparent or weakly absorbing layers, and significantly less visible in thick, highly absorbing layers.

2.6.2 Reflectance, Transmittance, and Absorption in Multilayer Stacks

Once the total transfer matrix is calculated, the reflection coefficient r is found using the same expression introduced earlier (in the lossless section). The reflectance is then

$$R(\lambda) = |r(\lambda)|^2 \tag{2.29}$$

If a transmission coefficient $t(\lambda)$ is also computed, the transmittance at normal incidence can be written in the standard form

$$\mathcal{T}(\lambda) = \frac{n_{\text{out}}}{n_{\text{in}}} |t(\lambda)|^2 \tag{2.30}$$

where n_{in} and n_{out} refer to the refractive indices of the incident and exit semi-infinite media, respectively (e.g. air on the front and glass or substrate on the back).

In many PV devices, the absorber layer is designed to be optically thick over most of the usable spectrum, meaning transmission becomes small. In that case, reducing reflection immediately increases absorption, which is why ARCs remain important even when absorption is strong.

2.7 Chapter Summary

1) Solar radiation spans UV, visible, and IR wavelengths, and atmospheric effects shape the terrestrial spectrum (AM 1.5G and AM 1.5D). PV current generation

depends on photon number, so irradiance is often converted into photon flux density.

2) The relationship between photon energy and wavelength explains why only photons with energy above the semiconductor bandgap contribute to useful carrier generation, while long-wavelength photons transmit and high-energy photons lose excess energy by thermalization.
3) QE connects optical input to electrical output.
   - EQE includes both optical and electrical losses
   - IQE removes reflection losses and focuses on carrier collection quality
   - QE spectra help diagnose where losses occur in a solar cell.
4) The photogenerated current density is computed from the weighted overlap of photon flux and EQE across wavelength. This links the device's optical design directly to the achievable short-circuit current.

$$J_{ph} = q \int_{\lambda_{\min}}^{\lambda_{\max}} \Phi(\lambda)\, \mathrm{EQE}(\lambda)\, d\lambda$$

5) Optical behavior inside materials is described by the complex refractive index

$$\tilde{n} = n_s - j\kappa$$

   where n_s is refractive index and κ is extinction coefficient, which is often expressed as the absorption coefficient α and absorption depth, δ.
6) Optical power is divided into reflectance (R), transmittance ($\mathscr{T}$), and absorption (A), satisfying conservation of energy:

$$R(\lambda) + \mathscr{T}(\lambda) + A(\lambda) = 1$$

   These quantities determine how many photons are available for carrier generation.
7) Reflection at a single interface is described using Fresnel coefficients, explaining why bare silicon reflects a large fraction of light due to its high refractive index mismatch with air.
8) Multilayer stacks require interference-based modeling. The TMM represents each layer using propagation and interface matrices, allowing efficient calculation of total reflectance and transmission for complex thin-film systems.
9) A quarter-wavelength ARC reduces reflection near a chosen design wavelength by forcing destructive interference between reflected waves. The best single-layer ARC performance also depends on index matching.
10) Including absorption in TMM uses the same framework but with a complex refractive index, so wave propagation includes both phase shift and amplitude decay. This enables modeling of realistic PV stacks where absorption is essential.

End-of-Chapter Problems

P2.1 Estimating J_{sc} from photon flux

Assume the solar photon flux is approximated by three wavelength bands given below. In each band, the photon flux density is taken as constant, and the QE is assumed constant. (Elementary charge: $q = 1.60 \times 10^{-19}$ C)

Three given bands:

1) 400 – 500 nm
   $\Phi_1 = 2.0 \times 10^{21}$ photons m^{-2} s^{-1} nm^{-1}
   $QE_1 = 0.85$
   $\Delta\lambda_1 = 100$ nm
2) 500 – 700 nm
   $\Phi_2 = 2.5 \times 10^{21}$ photons m^{-2} s^{-1} nm^{-1}
   $QE_2 = 0.90$
   $\Delta\lambda_2 = 200$ nm
3) 700 – 900 nm
   $\Phi_3 = 1.5 \times 10^{21}$ photons m^{-2} s^{-1} nm^{-1}
   $QE_3 = 0.60$
   $\Delta\lambda_3 = 200$ nm

   Find approximate short-circuit current density J_{sc}.

P2.2 Assume a uniform absorber where internal reflections are neglected (Beer–Lambert model). Assume absorption coefficient at a chosen wavelength to be $\alpha = 1.0 \times 10^4$ cm^{-1} and absorber thickness: $d = 2$ μm = 2×10^{-4} cm. Find transmitted intensity fraction, $\mathcal{T}$ (ignoring reflection).

P2.3 Using the same assumptions as Problem 2.2, estimate absorbed fraction A. Neglect reflection ($R = 0$).

P2.4 Consider a single absorbing film on a thick substrate, at one wavelength, under normal incidence. Suppose a TMM calculation gives the front-surface reflectance. Transmission through the film (into the substrate side) is then estimated using Beer–Lambert for attenuation in the film.

Given parameters are:

- TMM result at a given wavelength: $R = 0.08$
- Absorption coefficient of the film: $\alpha = 5.0 \times 10^3$ cm^{-1}
- Film thickness: $d = 3$ μm = 3×10^{-4} cm

Assume substrate is thick and any light reaching it is "transmitted" for the purpose of energy balance. Estimate transmittance, $\mathcal{T}$ and absorption, A.

P2.5 Compute how much of a monochromatic beam remains after passing through a 200 μm Si slab at wavelengths: 800 and 1100 nm. Ignore reflection.
Assume that: $\alpha(800\text{ nm}) = 10^3\text{ cm}^{-1}$ and $\alpha(1100\text{ nm}) = 10\text{ cm}^{-1}$

P2.6 Find the thickness of certain material needed to absorb 90% of the light (single pass, no reflections). Assume light wavelength of 800 nm and $\alpha = 10^3\text{ cm}^{-1}$.

P2.7 An ARC has $n_1 = 2.0$ and thickness $d = 80$ nm. What design wavelength λ_0 is targeted if quarter-wavelength concept was applied?

P2.8 ITO has $\kappa = 0.05$ at $\lambda = 500$ nm. Find α (use λ in meters).

P2.9 A bare silicon surface reflects strongly in the blue region. To reduce reflection at a design wavelength $\lambda_0 = 400$ nm, a single layer of indium tin oxide (ITO) is deposited on silicon. Assume all materials are lossless at λ_0, normal incidence and:

$$n_0 = 1.00(\text{air}), \quad n_1 = 1.90(\text{ITO at 400 nm}), \quad n_2 = 3.50\ (\text{Si at 400 nm})$$

a) Find the quarter-wave thickness d of the ITO layer at λ_0.
b) Using the single-layer thin film (equivalently TMM) result at normal incidence, compute the reflectance R at λ_0 for this quarter-wave layer.
c) Compute the bare silicon reflectance at λ_0 (air–Si interface).

Multiple Choice Question Bank

P2.10 Which relation correctly connects photon energy $h\nu$, wavelength λ, and frequency ν?

A $h\nu = \dfrac{\lambda}{hc}$

B $h\nu = \dfrac{hc}{\lambda}$

C $\dfrac{h}{\nu} = \dfrac{hc}{\lambda}$

D $h\lambda = \dfrac{\nu}{c}$

P2.11 Shorter-wavelength light corresponds to:

A Lower photon energy and lower frequency

B Higher photon energy and higher frequency

C Higher photon energy and lower frequency
D Lower photon energy and higher frequency

P2.12 Which photons can be absorbed in a semiconductor to generate electron–hole pairs?
A Only photons with $h\nu < E_g$
B Only photons with $h\nu > E_g$
C Only photons with $h\nu = 0$
D All photons, regardless of energy

P2.13 The solar spectrum standard AM0 corresponds to:
A Sunlight after passing through the atmosphere at $\theta = 48.2°$
B Sunlight measured at Earth's surface under cloudy conditions
C Sunlight outside Earth's atmosphere
D Sunlight reflected from Earth's surface

P2.14 Photon flux density $\Phi(\lambda)$ is defined as:
A Energy per photon per unit wavelength
B Number of photons per unit area per unit time per unit wavelength
C Optical intensity per unit area
D Absorbed photons per unit time

P2.15 External quantum efficiency EQE(λ) is defined as:
A Collected electrons/absorbed photons
B Reflected photons/incident photons
C Collected electrons/incident photons
D Absorbed photons/incident photons

P2.16 Internal quantum efficiency IQE(λ) differs from EQE because it:
A Includes reflection losses
B Excludes reflection losses
C Ignores recombination losses
D Is independent of wavelength

P2.17 In the EQE curve of a solar cell, reduced EQE at long wavelengths is often caused by:
A Strong absorption very close to the surface only
B Excess thermalization of carriers
C Insufficient absorption or insufficient thickness
D Perfect index matching at the surface

P2.18 The extinction coefficient κ is associated with:
A Phase velocity only
B Optical absorption (attenuation)
C Reflection at interfaces only
D Photon energy in vacuum

P2.19 If $n_1 = n_2$, then:
A $r_{12} = 1$ and reflection is maximum
B $r_{12} = 0$ and reflection is negligible
C r_{12} is imaginary only
D Reflection depends only on wavelength

P2.20 Why is bare silicon highly reflective in air?
A Silicon has $n \approx 1$, so it matches air
B Silicon has strong absorption only
C Large refractive index mismatch between air and silicon
D Silicon has zero extinction coefficient

P2.21 In multilayer optics, the electric field in any uniform layer can be written as:
A A single forward wave only
B A single backward wave only
C A sum of forward and backward-traveling waves
D A standing wave only (no traveling waves)

P2.22 The key advantage of TMM for multilayers is that it:
A Eliminates interference entirely
B Avoids wavelength dependence
C Accounts for multiple reflections and interference via matrix multiplication
D Works only for single interfaces

P2.23 A single-layer ARC reduces reflection mainly by:
A Increasing absorption in the coating
B Destructive interference between reflected waves
C Blocking photons below the bandgap
D Increasing the solar irradiance

P2.24 Interference effects are typically most pronounced in:
A Transparent or weakly absorbing layers
B Very thick, highly absorbing layers only

C Metals only
D Layers with $n = 1$ only

P2.25 Photon frequency at $\lambda = 600$ nm is closest to:
A 5.0×10^{14} Hz
B 2.0×10^{8} Hz
C 1.8×10^{-15} Hz
D 3.6×10^{20} Hz

P2.26 Photon flux density from irradiance (single wavelength) is given to be $S(\lambda = 600\text{ nm}) = 1.5\text{ W m}^{-2}\text{ nm}^{-1}$. Find $\Phi(\lambda)$.
A 4.5×10^{18}
B 4.5×10^{10}
C 1.5×10^{24}
D $2.0 \times 10^{15}\text{ s}^{-1}\text{ nm}^{-1}$

P2.27 A cell has EQE $= 0.80$ over a narrow band of $\Delta\lambda = 10$ nm . If photon flux density is constant over this band, $\Phi(\lambda) = 3 \times 10^{18}$ photons $\cdot\text{ m}^{-2}\text{ s}^{-1}\text{ nm}^{-1}$, find J_{ph} generated by this incoming light.
A 0.038 A/m^2
B 0.384 A/m^2
C 3.84 A/m^2
D 38.4 A/m^2

P2.28 At $\lambda = 500$ nm, $\Phi(\lambda) = 2 \times 10^{18}$ photons $\cdot\text{ m}^{-2}\text{ s}^{-1}\text{ nm}^{-1}$. Within $\Delta\lambda = 5$ nm, the measured current density contribution is $J_{ph} = 0.64$ A/m^2. Estimate EQE.
A 0.20
B 0.40
C 0.80
D 1.60

P2.29 At a wavelength, measured $R = 0.12$ and $\mathcal{T} = 0.70$. Compute absorption A.
A 0.18
B 0.42
C 0.82
D 1.00

P2.30 Air $n_1 = 1.0$, silicon $n_2 = 3.5$ (assume real). Find R.
A $R \approx 0.04$
B $R \approx 0.31$

C $R \approx 0.50$
D $R \approx 0.80$

P2.31 An ARC works mainly because it:
A Converts photons into heat before they reach silicon
B Creates destructive interference of reflected waves
C Increases κ to absorb all photons at the surface
D Makes $n = 1$ everywhere

P2.32 Air $n_0 = 1.0$, silicon $n_s = 3.5$. Find the ideal ARC refractive index.
A $n_1 = 1.2$
B $n_1 = 1.87$
C $n_1 = 2.50$
D $n_1 = 3.50$

P2.33 Measured EQE = 0.72, reflectance $R = 0.20$, assume $\mathcal{T} \approx 0$. Estimate IQE.
A 0.58
B 0.72
C 0.90
D 1.20

References

1 Sze, S.M. and Ng, K.K. (2007). *Physics of Semiconductor Devices*, 3rde. Hoboken: John Wiley & Sons.
2 ASTM G173-03. "Reference solar spectral irradiance: air mass 0, air mass 1.5 global, and air mass 1.5 direct. ASTM International. Data distributed by the National Laboratory of The Rockies (NRL)," 2025. [Online]. Available: https://www.nlr.gov/grid/solar-resource/spectra-am1.5. [Accessed 2025].
3 S. Abdul Hadi, *III–V on Si Multijuction Step-Cell: PhD Dissertation*, Abu Dhabi: Masdar Institute of Science and Technology (now Khalifa Unviersity), 2016.
4 Refractiveindex.info, "Refractive index database: optical properties of inorganic materials," 2026. [Online]. Available: https://refractiveindex.info/?shelf=main&book=MoS2&page=Islam-o. [Accessed 2026].

Reading Material

1 Nelson, J. (2010). *The Physics of Solar Cells*. London: Imperial College Press.
2 Gueymard, C.A., Myers, D., and Emery, K. (2002). Proposed reference irradiance spectra for solar energy systems testing. *Solar Energy* 73 (6): 443–467.

3

Physics of Semiconductors

3.1 About the Chapter

In order to understand how a solar cell works, it is vital to first understand the physics of semiconductors. Semiconductor materials are the building block of solar cell design. They are used to convert energy from the sun into electrical energy. While the Sun provides Earth with a constant energy source in the form of photons, this optical energy needs to be converted into electrical energy to power our devices. Thinking anecdotally, it would be nice and convenient to be able to charge a device, such as a cell phone, simply by placing it in sunlight. However, this will not work as there needs to be a material or device that can absorb the photons and convert them to electrons. This is where semiconductors are used since they can absorb photons that free electrons from their bound state. This is why semiconductors are used to make solar cells. This chapter begins with a brief overview of the crystal structure of semiconductors and how it leads to the formation of energy bands and the bandgap. After that follows a discussion on how bound electrons can become charge carriers in semiconductors. This leads to the concept of a positively charged hole as a charge carrier. Statistics are then used to calculate the number of electrons and holes in the various bands, and we explain how the charge can be increased. Finally, we examine the physics of how charge carriers move under an electric field and by diffusion, giving rise to drift and diffusion currents.

Key Topics

- Energy bands and bandgap formation
- Metals, insulators, and semiconductors
- Electrons, holes, and carrier generation
- Doping and Fermi level
- Drift, diffusion, and mobility
- Excess carriers and recombination

3.2 Energy Bands

3.2.1 Bandgap

To understand the physics of semiconductors, it is important to understand the basic principles of energy bands. More importantly, how do semiconductor crystals form, and what happens at the atomic level. The first fundamental discovery of quantum mechanics is that the energy of the bound electron is quantized. This means that electrons can only have discrete energy levels. The other fundamental point is that the electron is not localized at a given radius. This is the wave–particle duality that states that electrons can exhibit both particle and wavelike properties. The Nobel Prize in Physics in 1929 was awarded to Prince Victor Pierre Raymond de Broglie "for his discovery of the wave nature of electrons" highlighting the significance of the discovery. Using these two principles, crystal formation can be explained by introducing forbidden energy bands and bandgap. First, consider two isolated atoms and bring them closer together. In this system, as the crystal is forming, there exists an interatomic spacing that balances all the forces that interact. Figure 3.1 shows energy vs. interatomic spacing in Si.

In silicon, the valence and conduction bands originate from the splitting and hybridization of the atomic 3s and 3p orbitals. Bonding combinations form the valence band, while antibonding combinations give rise to the conduction band. During crystal formation, changes occur in the electron energy levels, which

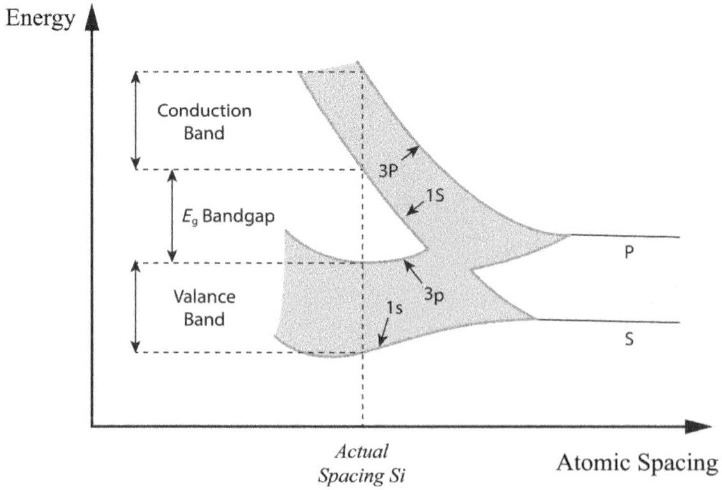

Figure 3.1 Energy vs. atomic spacing in silicon.

Figure 3.2 Simplified Si bandgap model.

explain the differences in the electrical and optical properties of materials. As the atomic spacing is reduced and the crystal forms, discrete atomic energy levels split due to interactions between electron wavefunctions until an optimal interatomic spacing is reached. At this equilibrium spacing, a valence band and a conduction band are formed, separated by a forbidden energy bandgap. Every material has a unique energy separation between the valence and conduction bands, known as the energy bandgap E_g. The Pauli exclusion principle states that no two electrons in the same atom can have identical quantum numbers or occupy the same quantum state; consequently, electrons within a band are restricted to distinct energy levels.

Figure 3.2 shows a simplified bandgap model for Si. E_c is the conduction band, E_v is the valence band, and E_g is the bandgap. Note that in the simplified model, E_c and E_v are drawn as single lines, but it is important to remember it is a parabolic band with levels as function of energy, as shown in Figure 3.1.

3.2.2 Metals, Insulators, and Semiconductors

This size of bandgap allows for sorting and classification of materials into three distinct categories. The first category is when the energy bands overlap, so essentially, there is no bandgap. In these materials, electrons move freely without additional energy. These are called conductors since they conduct electricity easily, such as metals like gold, silver, or aluminum. On the other hand, if the bandgap is very large, it becomes virtually impossible for an electron to make it to the conduction band. In this case, the material does not conduct electricity and is an insulator. Some everyday examples of insulators are wood, glass, or a piece of cloth. The third type of materials has a bandgap smaller than that of insulators. This means the energy needed to excite an electron from the valence band to conduction band is not very large and is easily achievable. These materials are known as semiconductors and, as the name implies, their conductivity can be controlled. This feature allows for modification of the material properties for various applications. Semiconductors are used to make various important devices such as diodes, transistors, or solar cells. Some examples include silicon, germanium (Ge), and carbon from group IV of the periodic table. The bandgap of semiconductors can range from 0.6 to 3 eV, while insulators have bandgap larger

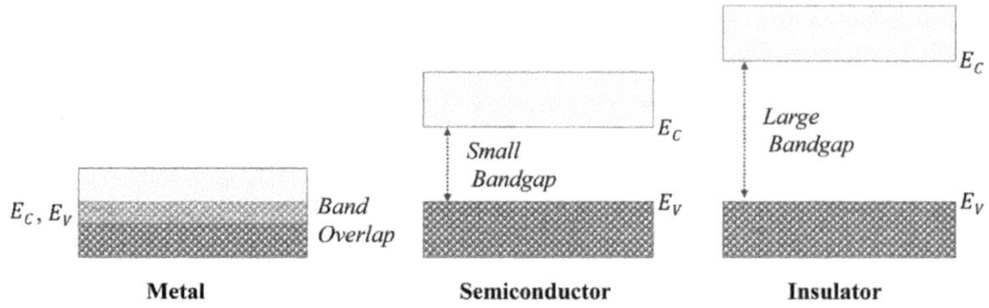

Figure 3.3 Illustration of band diagram of metals, insulators, and semiconductors.

than 3 eV. Figure 3.3 shows the simplified bandgap diagrams of the three types of materials.

3.3 Charge Carriers in Semiconductors

3.3.1 Electron–Hole Pairs

How do bound electrons make it from valence band into the conduction band? We can use Si as an example to understand the mechanism and physics of this process. Figure 3.4 shows a schematic of crystal structure of Si. An individual Si atom is surrounded by four other Si atoms bonded covalently (shared electron). When energy is absorbed in Si, the covalent bond can break, freeing up an electron to transition to the conduction band. This is possible since in a covalent bond, the electron is bound to the bond and not the atom, so the binding energy is lower. As the thermal energy is increased, this energy is used to break the bond and free the electron.

In the valence band, the missing electron leaves an empty bond behind that we define as a hole. While physically a hole is simply the absence of an electron, mathematically speaking, a hole can be treated as a positively charged particle. However, it is important to keep in mind that a hole is a missing bond and not an

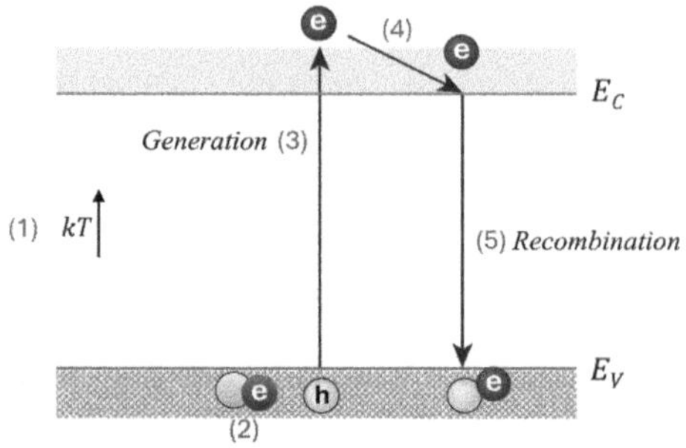

Figure 3.4 Band diagram of generation recombination process in equilibrium due to thermal energy.

actual charged particle. This mathematical construct helps to quantify the amount of charge carriers in the semiconductor. So let us define n for electron concentration (negative charge) and p for hole concentration (positive charge). So, in a pristine semiconductor with no external elements added, for every electron that makes it to the conduction band, a hole is also created, and these are called electron–hole pairs (EHPs) (Eq. (3.1)). Note the units for concentration in semiconductor are per unit volume, cm^{-3}.

$$n = p = \text{EHPs}\,(\text{cm}^{-3}) \tag{3.1}$$

As the thermal energy increases, more EHPs are formed. To understand this, let us breakdown the process in steps: (1) thermal energy absorbed, (2) covalent bond is broken and electron is free from its bound state, hole is formed (EHP), (3) electron gains energy equal to or larger than E_g and makes it to the conduction band (*generation*), (4) the electron loses energy (*more info on this later*), and (5) electron returns to the valence band and fills the empty hole (*recombination*). Figure 3.4 shows a schematic of the process from steps (1) to (5).

3.3.2 Generation/Recombination Rates—Intrinsic Concentration

As thermal energy increases, more covalent bonds are broken, allowing additional electrons to make it to the conduction band and holes to be created in the valence band. We can define G as the generation rate and $\mathcal{R}$ as the recombination rate with units of $\text{EHPs/cm}^3\cdot\text{s}$. This is the number of EHPs created or recombined per unit volume per second. In thermal equilibrium, there are certain number of electrons in the conduction band and a certain number of holes in the valence band at any given time. Thermal equilibrium means, at that particular temperature, the rate of generation of electrons and holes equals the rate of recombination of electrons and holes (Eq. (3.2)).

$$G = \mathcal{R} \tag{3.2}$$

One way to help visualize this is to take a screenshot (after some time) and examine both bands. This is called thermal equilibrium. The number of carriers in this screenshot is called the intrinsic carrier concentration, n_i, with units cm^{-3} (Eq. (3.3)). In thermal equilibrium, the concentration of holes equals the concentration of electrons; thus, the following relationship holds

$$n = p = n_i \tag{3.3}$$

Also, we can use the following simple relationship for easy calculations:

$$np = n_i^2 \tag{3.4}$$

Each semiconductor will have an intrinsic carrier concentration at room temperature based on the bandgap size that was formed during the crystal formation.

Table 3.1 Key semiconductor materials (Si, Ge, and GaAs), bandgap value, and n_i value (300 K).

Material	E_g (eV)	n_i (cm^{-3})	B (cm^{-3} $K^{-3/2}$)
Si	1.10	1.5×10^{10}	5.23×10^{15}
Ge	0.67	2.5×10^{13}	1.66×10^{15}
GaAs	1.43	2.0×10^{6}	2.10×10^{14}

The smaller the bandgap, the higher the intrinsic concentration, since less energy is needed for the EHPs to form. For example, n_i for Si at room temperature is 1.5×10^{10} cm^{-3} and for Ge it is 2.4×10^{13} cm^{-3}. Intrinsic concentration will also increase with increasing temperature due to higher thermal energy. Intrinsic concentration, n_i, can be calculated based on Eq. (3.5).

$$n_i(T) = BT^{3/2}e^{-E_g/2kT} \tag{3.5}$$

where T is temperature, E_g is material-specific bandgap, and k is Boltzmann constant. The constant B is material-specific and it is calculated from effective density of states in valence (N_v) and conduction (N_c) bands ($B = \sqrt{N_c N_v}$). Table 3.1 shows the bandgap values and intrinsic concentration values at room temperature for semiconductor materials (Si, Ge, and GaAs).

When a semiconductor has equal concentration of electrons and holes, and is in pristine form, it is called *intrinsic*. This is the semiconductor that is right "out-of-the-box" with no external impurities added. Figure 3.5 plots n_i vs. temperature for Si, Ge, and GaAs. As shown, the intrinsic concentration can become very large at high temperatures. For example, n_i for Si at 500 K is 2×10^{14} cm^{-3}. The plot also highlights the significance of the E_g size on the n_i value.

3.3.3 Extrinsic Semiconductor and Doping

Most devices can not operate with low intrinsic carrier concentration since they will not be very conductive. However, what makes semiconductors special compared to other materials is that the number of electrons and holes in the conduction and valence bands can be increased by adding external atoms to the crystal. This process is called doping, and external atoms are often referred to as impurities or dopants. The model of doping is that the external atoms (dopants) take the place of the Si atom in the crystal, trying to fulfil its bonding requirements. To dope Si (group IV), group III or V of elements are used. If Si is doped with a group III element, such as boron (B) or aluminum (Al), a hole is added since group III atoms

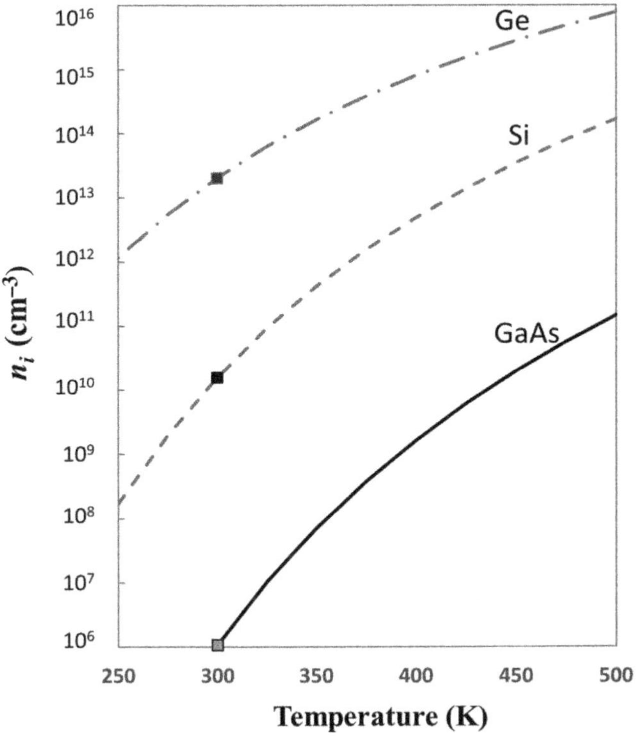

Figure 3.5 n_i as a function of temperature T (K) for Ge, Si, and GaAs.

have one less electron in the outer shell compared to Si, leaving an empty bond. This extra bond adds a hole to the valence band of the Si, increasing the hole concentration. These dopants from group III are called *acceptors*. On the other hand, if group V atoms, like P or As, are added to Si, there will be an extra electron in the outer shell that is loosely bound to the atom after all four of the Si covalent bonds are completed. This extra electron is added to the conduction band of the Si, increasing the electron concentration. Thus, atoms from group V are referred to as *donors*. Figure 3.6 is a schematic model of the doping process.

A band diagram model is used to understand how the doping atom can provide additional carriers is shown in Figure 3.7. The donor state E_d will be near the conduction band and the acceptor state E_a is near the valence band. In this case, only a small amount of thermal energy is needed for an electron to move from donor state E_d to the conduction band. This is true since the extra electron is loosely bound to the dopant atom not to the covalent bond. Similarly, very little energy is needed for the electron to go from valence band to acceptor state (E_a). This implies

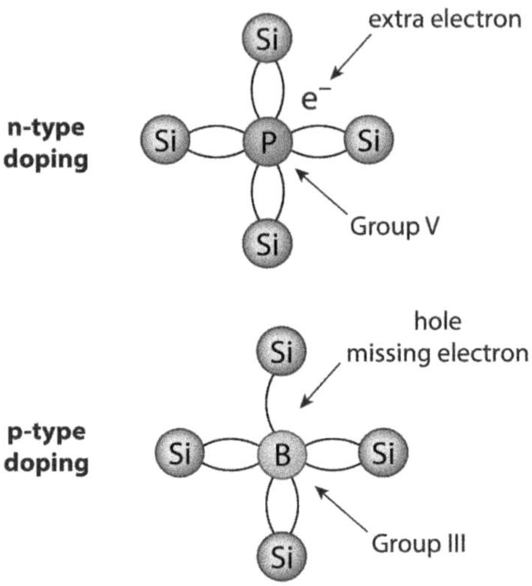

Figure 3.6 Schematic model of *p*- and *n*-type doping in silicon.

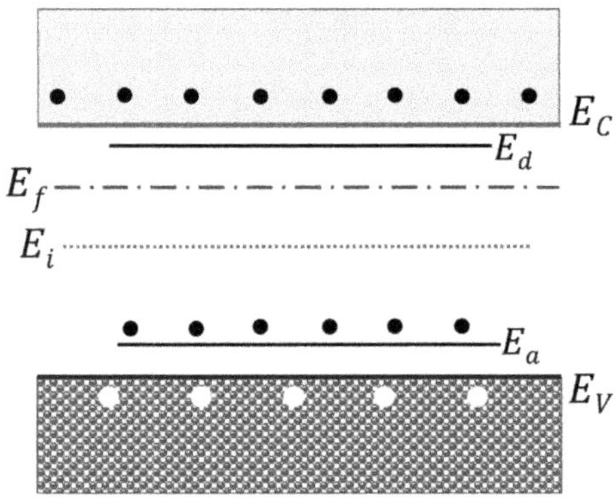

Figure 3.7 Band diagram model for *n*-type semiconductor showing the doping levels.

that if a semiconductor is doped with a significant number of donors, for example, $N_d \gg n_i$, then

$$n_0 \cong N_d \tag{3.6}$$

If the semiconductor has a larger number of acceptors, $N_a \gg n_i$, then

$$p_0 \cong N_a \tag{3.7}$$

3.4 Carrier Concentrations

3.4.1 Fermi–Dirac Statistics

To understand the behavior of semiconductors, it is important to calculate how many electrons and holes are in the various bands. To quantify this, statistics are needed due to the large number of available states and electrons. Statistics can be used since all electrons share the same physical properties and are therefore mutually indistinguishable in all respects. The very same holds for all protons, neutrons, and other particles of the same kind in the universe. Statistics can tell us the probability that an electron is located at a certain energy in the band structure of semiconductor. To do this, we can use Fermi–Dirac statistics. It is named after Enrico Fermi and Paul Dirac, each of whom derived the distribution independently in 1926. Fermi–Dirac statistics is a type of quantum statistics that applies to the physics of a system consisting of many non-interacting, identical particles that obey the Pauli exclusion principle. The result is the Fermi–Dirac distribution of particles over energy states. The following formula is the Fermi–Dirac function:

$$f_F(E) = \frac{1}{1 + e^{(E-E_f)/kT}} \tag{3.8}$$

where E_f represents the Fermi level and is defined as the energy level where the probability of an electron being located is $\frac{1}{2}$. The Fermi level, E_f, is one of the most important concepts in semiconductor device physics as it represents a simple way to view entire Fermi function. Using Eq. (3.8), the probability of electron occupancy vs. energy for different temperatures is plotted in Figure 3.8.

At 0 K, the valence band is full of bound electrons, and the conduction band is empty. So, the probability that the electron is above E_f is 0 and probability that an electron is below E_f is 1 resulting in a step function, as shown in Figure 3.8. As the temperature starts to increase, covalent bonds break and EHPs are formed. Electrons now have enough energy to make it to the conduction band. In this case, the probability that an electron is above the E_f is non-zero. As the temperature continues to increase, more EHPs are formed, and as a result, more electrons

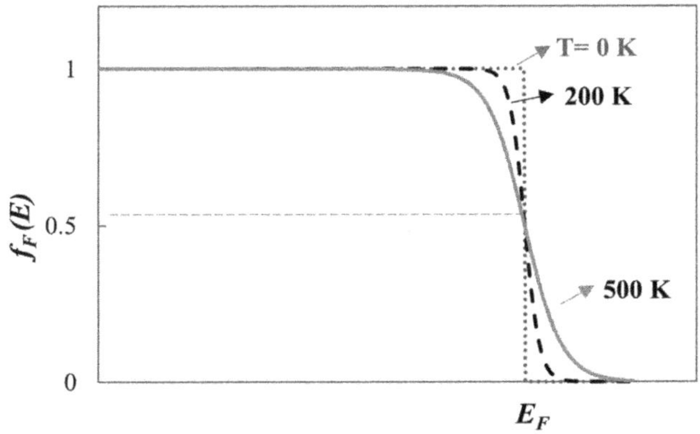

Figure 3.8 Probability of electron occupancy, $f_F(E)$, vs. energy, E, at different temperatures.

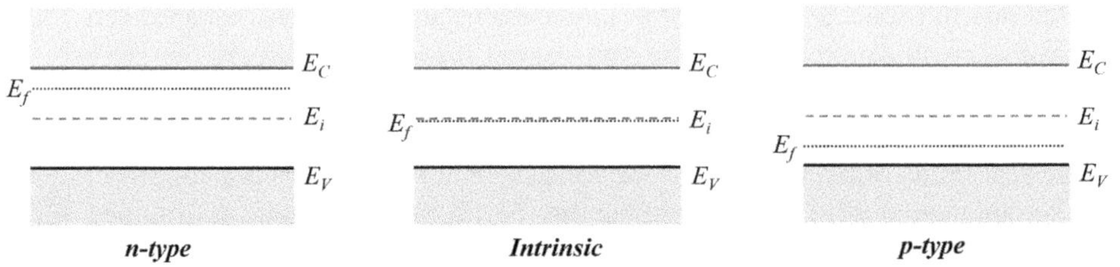

Figure 3.9 Simplified band diagram illustration for *n*- and *p*-type semiconductors with Fermi levels indicated.

make it to the conduction band. This means that the probability of an electron is above E_f continues to rise as shown in Figure 3.8.

Since it is difficult to plot and draw the Fermi probability function for each case, position of E_f within the bandgap is used. If a semiconductor has more electrons than holes, it is called *n*-type and E_f will be closer to conduction band energy, E_c. If a semiconductor has more holes than electrons, it is called *p*-type and E_f is closer to valence band energy, E_v. As the *n*-type doping is increased E_f will move closer to E_c. Similarly, as the *p*-type doping increases, E_f moves closer to E_v. If the semiconductor has an equal number of holes and electrons, such as in intrinsic materials, then E_f will be in the middle of the bandgap. Figure 3.9 shows a simplified band diagram with E_f location for *n*-type, intrinsic, and *p*-type materials.

3.4.2 Energy Bands and Density of States

In the conduction and valence bands, there is an available density of states for electrons/holes to reside in. Using this, combined with the Fermi function, we can calculate the total number of electrons in the conduction band and holes in the valence band. The probability that an electron or hole occupies a certain energy state multiplied by the number of available states at that energy will tell us the concentration of the electrons. Then we can integrate from the edge of the conduction band to infinity to add up electrons in the conduction band. The electron concentration is defined as Eq. (3.9):

$$n(E) = \int_{E_c}^{\infty} f_F(E)\, N(E)\, dE \tag{3.9}$$

The total number of available states can be modeled to be residing at the conduction band E_c, so the solution is Eq. (3.10).

$$n(E_c) = N_c f_F(E_c) \tag{3.10}$$

where N_c is defined as the density of states in the conduction band. Given that the value of kT at room temperature is 0.0259 eV and E_f is several kT below the conduction band edge (E_c), the exponential term in the Fermi–Dirac distribution becomes much larger than unity. Under this condition, the Fermi–Dirac distribution can be approximated by the *Maxwell–Boltzmann distribution*. Thus, the electron occupancy probability at the conduction band edge may be written as

$$f_F(E_c) = \frac{1}{1 + e^{(E_c - E_f)/kT}} \approx e^{-(E_c - E_f)/kT} \tag{3.11}$$

So, the total amount of electrons in the conduction band n_0 is defined by

$$n_0 = N_c\, e^{-(E_f - E_c)/kT} \tag{3.12}$$

We can follow the same analysis for holes in valence band. Using the Fermi–Dirac function with *Maxwell–Boltzmann distribution* the probability of hole occupancy at E_v is given as

$$f_F(E_v) = \frac{1}{1 + e^{(E_v - E_f)/kT}} \approx e^{-(E_f - E_v)/kT} \tag{3.13}$$

The concentration of holes in valence band is the integral across the valence band of probability that a hole is located at a certain energy level multiplied by number of states at that level and is described as

$$p = \int_{E_v}^{\infty} f_F(E)\, N(E)\, dE \tag{3.14}$$

The solution to this integral is the total hole concentration in the valence band, which is defined as

$$p_0 = N_v\, e^{-(E_f - E_v)/kT} \tag{3.15}$$

where N_v is the density for states in the valence band. For an intrinsic material, the E_f will lie in the middle of the bandgap (E_i) as shown in Figure 3.10.

The product of n_0 and p_0

$$n_0 p_0 = \left(N_c\, e^{-(E_f - E_c)/kT}\right)\, \left(N_v\, e^{-(E_f - E_v)/kT}\right)$$

$$n_0 p_0 = N_c N_v e^{-(E_c - E_v)/kT}$$

Since $E_c - E_v = E_g$, then

$$n_0 p_0 = N_c N_v e^{-(E_g/kT)} \tag{3.16}$$

One interesting and important ramification of Eq. (3.16) is that the product of n_0 and p_0 is constant.

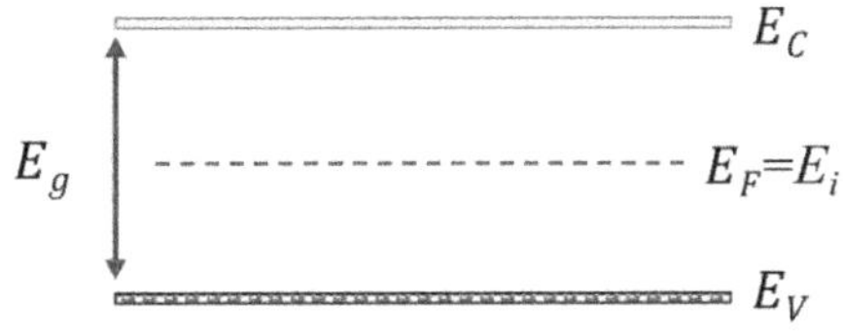

Figure 3.10 Simplified band diagram showing the intrinsic level at the midgap.

3.4.3 Calculation of the Intrinsic Concentration

Using Eq. (3.16), a formula for intrinsic carrier concentration can be derived. At thermal equilibrium, the number of electrons equal the number of holes $n_0 = p_0$ and the concentration is the intrinsic concentration n_i. Since $n_0 = p_0 = n_i$, then

$$n_i^2 = N_c N_v e^{-(E_g/kT)}$$

and

$$n_i = \sqrt{N_c N_v e^{-(E_g/kT)}} \tag{3.17}$$

Also,

$$n_0 p_0 = n_i^2 \tag{3.18}$$

One conclusion from Eq. (3.17) is that n_i is strong function of the material bandgap E_g and the temperature T, since it is part of the exponential term.

The formula for n_0 and p_0 are currently written in terms of N_v, N_c, and E_f relative to E_c and E_v. However, new formulas for n_0 and p_0 as a function of n_i and E_f relative to E_i can be derived. Since a semiconductor is intrinsic that means E_f is in the middle of the bandgap as shown in Figure 3.10. So, setting $E_f = E_i$ and $n_0 = n_i$ Eq. (3.12) can be written as

$$n_i = N_c e^{-(E_c - E_i)/kT} \tag{3.19}$$

By using some mathematical manipulation of the exponential term of Eq. (3.12) yields

$$n_0 = N_c e^{(-(E_c - E_i) + (E_f - E_i))/kT}$$

and

$$n_0 = N_c e^{-(E_c - E_i)/kT} e^{(E_f - E_i)/kT}$$

Using Eq. (3.19), then

$$n_0 = n_i e^{(E_f - E_i)/kT} \tag{3.20}$$

This derivation can also be done for holes in the valence band. In an intrinsic semiconductor when $p_0 = p_i = n_i$ then $E_f = E_i$ near half the bandgap. So, we can derive

$$p_i = n_i = N_v e^{-(E_i - E_v)/kT} \tag{3.21}$$

Using some mathematical manipulation of the exponential term of Eq. (3.21) yields

$$p_0 = N_v e^{(-(E_i - E_v) + (E_i - E_f))/kT}$$

and

$$p_0 = N_v e^{-(E_i - E_v)/kT} e^{(E_i - E_f)/kT}$$

Using Eq. (3.21), we get that

$$p_0 = p_i e^{(E_i - E_f)/kT} = n_i e^{(E_i - E_f)/kT} \tag{3.22}$$

Equation (3.22) represents the total hole concentration as a function of n_i and the energy separation between $E_i - E_v$.

3.4.4 Compensation and Total Concentration as a Function of Temperature

The concept of doping was introduced earlier in the chapter. Doping was modeled as an energy state near the conduction or valence band. Only a small amount of energy is needed to ionize the dopant. What about the case when semiconductor is doped with donor and acceptor doping at the same time. Intuitively, the doping type that has higher amount will dominate. Mathematically using charge neutrality the total positive charge must equal the total negative charge. The positive charges are from the mobile holes, and the ionized donor doping atoms. The negative charge is the mobile electrons and the ionized acceptor atoms. So, to maintain charge neutrality, we get

$$p_0 + N_d^+ = n_0 + N_a^- \tag{3.23}$$

If we solve for the total electron concentration, we obtain

$$n_0 = p_0 + \left(N_a^- - N_d^+\right) \tag{3.24}$$

and if all the doping atoms are ionized, then

$$N_d^+ - N_a^- = N_d - N_a \tag{3.25}$$

Further, if the material is n-type with $n_0 \gg p_0$, then

$$n_0 = N_d - N_a \tag{3.26}$$

Also, if the material is p-type and $p_0 \gg n_0$, then

$$p_0 = N_a - N_d \tag{3.27}$$

Using all these concepts and derivation, the total carrier concentration can be plotted as a function of temperature. At low temperatures the entire donor state can be ionized filling the conduction band with electrons. Let us define N_d as a donor doping concentration and N_a as the acceptor doping concentration and check holistically how the concentration of holes and electrons changes

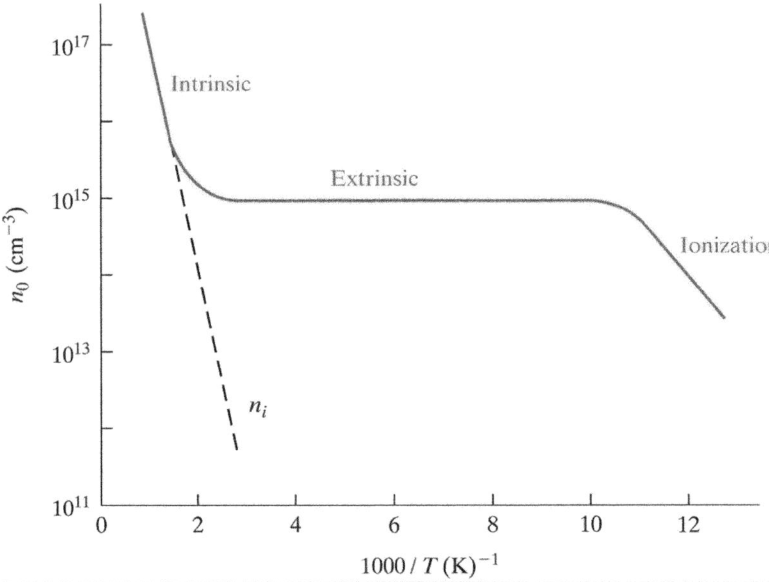

Figure 3.11 Carrier concentration vs. temperature. *Source:* [1] / with permission of Pearson Education.

as a function of temperature, considering all factors, both thermal and doping. Figure 3.11 illustrates electron and hole concentration vs. temperature for semiconductors with a doping concentration of $N_d = 1 \times 10^{15}$ cm^{-3} as a baseline.

At 0 K, or sometimes called freeze out, there is no thermal energy provided to break a covalent bond. As a result, no EHPs are created either from the doping state or from the valence band. As the temperature slowly increases, the number of carriers increases initially due to the doping atom. The doping atom level is very close to the conduction band, so very little energy is needed. This makes sense since the electron is loosely bound to the doping atom compared to tightly bound in the covalent bond. As the temperature increases, more carriers are ionized from the donor state until eventually all carriers make it to the conduction band and the doping atom is fully ionized. At this point, any additional thermal energy will only increase the intrinsic concentration. Since the value of the doping is much larger than n_i ($N_d \gg n_i$) the increase in temperature does not significantly affect the overall number of free carriers. However, as the temperature continues to increase, n_i can reach very high values and become larger than the doping concentration ($n_i \gg N_d$). At this point, the material is dominated by the thermal generation n_i and is considered intrinsic despite the doping. This ability of electrons and holes to be added to the conduction and valence band is what makes semiconductors unique compared to insulators and conductors.

Example 3.1 ***Intrinsic concentration at different temperatures***
In a phosphorus-doped silicon layer with doping concentration of 10^{18} cm^{-3}, find the hole and electron concentrations at 137°C.

Solution
As phosphorus is n-type dopant, then $n \cong N_d = 10^{18}$ cm^{-3}

$$T = 137°\text{C} = 273 + 137 = 410\ \text{K}$$

At $T = 410$ K, $n_i = BT^{3/2}e^{-E_g/2kT} = 7.3 \times 10^{15} \times (410)^{3/2}e^{-1.12/(2\times8.62\times10^{-5}\times410)}$
$n_i = 7.96 \times 10^{12}$ cm^{-3}

Hole concentration in the n-type material is given by

$$p_n = \frac{n_i^2}{N_d} = \frac{(7.92 \times 10^{12})^2}{10^{18}} = 6.34 \times 10^7\ \text{cm}^{-3}$$

Example 3.2 ***Fermi level and electron hole concentration***
Suppose you have a silicon semiconductor doped with 10^{16} cm^{-3} Boron.

a) Calculate at room temperature (300 K) the electron and hole concentrations.
b) Calculate the $E_i - E_f$.
c) Draw and label the band diagram.

Solution
a) Looking at the periodic table, Boron is in group III. Thus, it is an acceptor, with doping $N_a = 10^{16}$ cm^{-3} and will provide additional holes. Since n_i for Si at room temperature is 1.5×10^{10} cm^{-3} the $N_a \gg n_i$.
Then

$$p_o \cong N_a = 1 \times 10^{16}\ \text{cm}^{-3}$$

Using $n_0\, p_0 = n_i^2$ it follows that

$$n_0 = \frac{n_i^2}{p_0}$$

Then,

$$n_0 = \frac{(1.5 \times 10^{10})^2}{10^{16}} = 2.25 \times 10^4\ \text{cm}^{-3}$$

b) Since the material is p-type the Fermi level will be below the E_i (closer to E_v than E_c). Using Eq. (3.22) and solving for $E_i - E_f$,

$$E_i - E_f = kT\ \ln\frac{p_0}{n_i} = 0.0259 \times \ln\frac{1 \times 10^{16}}{1.5 \times 10^{10}} = 0.347\ \text{eV}$$

c)

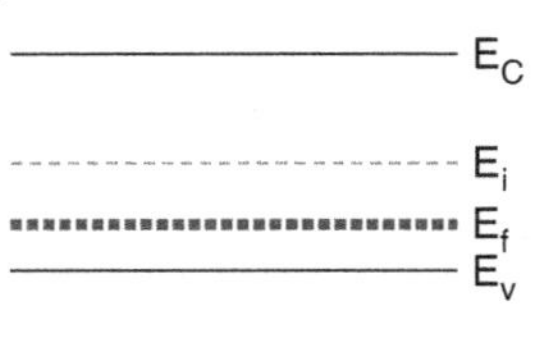

3.5 Carrier Drift

3.5.1 Drift Current

For a current to flow in a semiconductor, a density of charged carriers moving in one direction is needed. As explained earlier, a semiconductor at room temperature will have an intrinsic concentration of electrons in the conduction band and holes in the valence band. These free carriers, once in the conduction band, move due to random thermal motion, but not all in the same direction. During the random motion, the carriers can collide with each other, the lattice, or the ionized impurities, changing their direction. This process is called *scattering*. This random thermal motion of the charged particles does not result in a net current since they are moving in different directions.

So how can electrons or holes be guided to move in one direction? The answer is to use an electric field. In electromagnetism, the Lorentz force (or electromagnetic force), a particle of charge q moving with a velocity v in an electric field (E) will experience a force. So, if electrons or holes are exposed to an electric field, then a directed force is applied to them. Electrons, being negatively charged, move opposite to the electric field (toward the positive terminal), while holes, being positively charged, move in the direction of the electric field (toward the negative terminal). Remember that the electric field originates from positive charges and terminates at negative charges. The motion of charged particles due to an electric field is called *drift*. Figure 3.12a depicts the random thermal motion of an electron in the absence of an electric field. Figure 3.12b shows that under an electric field, the electron's motion is not random at the macroscale, although it remains random at the microscale. This is important because when many electrons drift in one direction, a net electric current is produced.

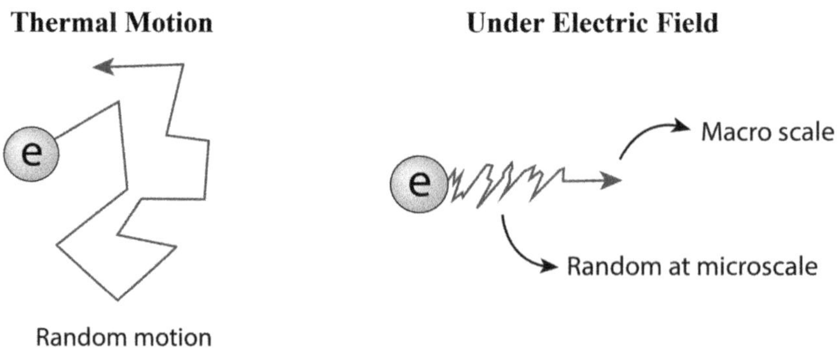

Figure 3.12 (a) Electron moving with random thermal motion no electric field and (b) electron in an electric field moving in one direction in micro scale.

Electrons moving with average velocity can produce a current density equal to

$$J_{drift} = -qn\bar{v}_x \tag{3.28}$$

with units

$$J_{drift} = (C)\left(\frac{1}{\text{cm}^3}\right)\left(\frac{\text{cm}}{\text{s}}\right) = \frac{C}{\text{cm}^2\,\text{s}} = \frac{A}{\text{cm}^2} \tag{3.29}$$

We can now relate Newton's second law that says that the acceleration of an object is directly proportional to the net force acting upon it and inversely proportional to its mass to the Lorentz force (F) for a charged particle under an applied electric field (E) using

$$F = qE = m^*a \tag{3.30}$$

As the electron moves in the crystal for example, it will increase its velocity, but the increase is not linear due to scattering. The electrons will be involved in collisions due to ionized impuriti and lattice atoms vibrating. When a collision happens, it will lose all its energy before accelerating again and experiencing another collision. This process is repeated so we can define an average velocity over many scattering events. This average velocity of electrons is directly proportional to the electric field in the x direction, E_x.

$$\overline{v_x} = -\mu_n E_x \tag{3.31}$$

where μ_n is defined as the electron mobility with units $\text{cm}^2/(\text{V}\cdot\text{s})$. Mobility is one of the most important parameters in semiconductor device physics and is a measure of how well charge carriers move in a crystal under electric field. We will discuss the physics of its properties in the next section. Continuing with the derivation, if we substitute Eq. (3.31) into Eq. (3.28), we get current density due to electron drift to be equal to

$$J_n(drift) = -qn\mu_n E_x \tag{3.32}$$

We can rewrite (3.32) like in Ohm's law format

$$J_n(drift) = \sigma_n E_x \tag{3.33}$$

and define the electron conductivity σ_n $(\Omega\ \text{cm})^{-1}$ in terms of mobility by

$$\sigma_n \equiv qn\mu_n \tag{3.34}$$

Note the reciprocal of the conductivity is resistivity, ρ (Ω cm). The same analysis can be done for holes moving in a semiconductor. A current density due to holes moving with average velocity is defined as

$$J_p(drift) = qp\bar{v}_x \tag{3.35}$$

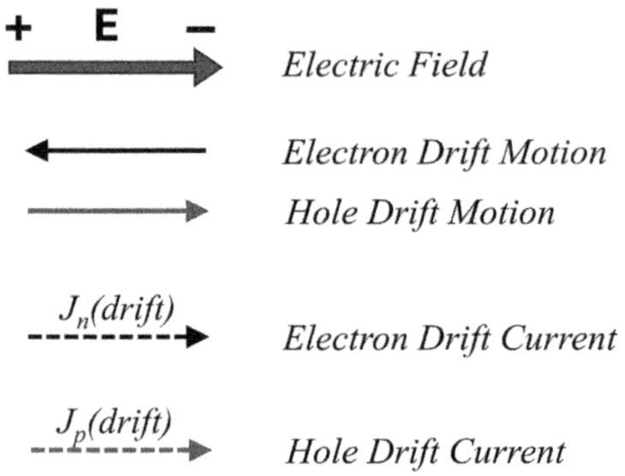

Figure 3.13 Motion of carriers and drift current under an electric field.

As in the electron case, the average velocity of the hole is directly proportional to the electric field by

$$\overline{v_x} = \mu_p E_x \tag{3.36}$$

where μ_p is defined as the hole mobility with units $\text{cm}^2/(\text{V}\cdot\text{s})$. If we substitute Eq. (3.36) into Eq. (3.35), we get

$$J_p(drift) = qn\mu_p E_x \tag{3.37}$$

We also rewrite Eq. (3.37) in Ohm's law format, and this gives

$$J_p(drift) = \sigma_p E_x \tag{3.38}$$

and define the hole conductivity σ_p $(\Omega\ \text{cm})^{-1}$ in terms of hole mobility.

$$\sigma_p \equiv qp\mu_p \tag{3.39}$$

Figure 3.13 shows the motion of both electrons and holes under an electric field and the corresponding current directions. Note that the current due to the electron motion is in the opposite direction of the motion of electrons since they are negatively charged.

3.5.2 Physics of Mobility

From Eq. (3.30), which relates the Lorentz force to Newton's second law, we can replace the acceleration with $\frac{dv}{dt}$ where $\overline{v}$ is the average velocity.

$$F = qE = m_n^* \frac{dv}{dt} \tag{3.40}$$

Since the electric field and effective mass are constant, we can integrate Eq. (3.40) taking the initial velocity of zero, so

$$v = \frac{qt}{m_n^*} E \tag{3.41}$$

Table 3.2 Bulk mobility values for Si, Ge, and GaAs materials.

	μ_n cm²/(V · s)	μ_p cm²/(V · s)
Si	1350	480
Ge	3900	1900
GaAs	8500	400

And substituting Eq. (3.41) into Eq. (3.37),

$$v = \frac{qt}{m_n^*}E = \mu_n E \tag{3.42}$$

$$\mu_n = \frac{q\bar{t}}{m_n^*} \tag{3.43}$$

where $\bar{t}$ is the time between collisions, and m_n^* is effective mass of the electron.

The same derivation for holes can be done, and the hole mobility with units $\mathrm{cm}^2/(\mathrm{V} \cdot \mathrm{s})$ is

$$\mu_p = \frac{q\bar{t}}{m_p^*} \tag{3.44}$$

Equations (3.44) and (3.43) highlight two key physical parameters that define the mobility: effective mass and the time between collisions. As a result, mobility can be increased by reducing the effective mass of the charged carrier or increasing the time between collisions. The effective mass can be modified by changing the semiconductor material used. Bulk mobility values of some common materials are listed in Table 3.2.

3.5.3 Total Drift Current

Note, the direction of current in the conventional sense is based on the positive charge carrier. In this case, the current will be in the direction of the hole drift. So, drift current due to holes under electric field is given by

$$J_p = qp\mu_p \mathrm{E} \tag{3.45}$$

and the drift current due to electrons is

$$J_n = qn\mu_n E \tag{3.46}$$

So, the total current is

$$J_{total} = J_n + J_p \tag{3.47}$$

The total current in terms of key parameters is

$$J_{total} = (n\mu_n + p\mu_p)qE = \sigma E \tag{3.48}$$

Equation (3.48) is the total drift current under electric field (E) from both electrons and holes with conductivity $\sigma = (n\mu_n + p\mu_p)q$.

3.5.4 Semiconductor as a Resistor

The semiconductor bar can also be thought of as a resistor. Let us consider a semiconductor bar connected to a battery with constant current flowing as shown in Figure 3.14.

The resistance of a cylindrical shape or a cuboid is calculated from the resistivity ρ as using

$$R = \rho \frac{L}{A} \tag{3.49}$$

where L is the length, and A is the cross-sectional area of the bar through which current flows. Substituting the material parameters for resistivity in Eq. (3.49), we obtain

$$R = \frac{L}{(n\mu_n + p\mu_p)qA} \tag{3.50}$$

From Eq. (3.50), the resistance of semiconductor bar can be calculated by knowing its doping and dimensions.

Example 3.3 ***Semiconductor as a resistor***

Given a Si bar, 2-cm-long with a cross-sectional area = 0.05 cm^2. Si bar is doped with $N_d = 10^{16}$ cm^{-3}. Assume that $\mu_p = 500$ cm^2/(V · s) and $\mu_n = 1400$ cm^2/(V · s) and find the following:

a) the concentration of electrons, n, and concentration of holes, p,
   the conductivity, σ,
b) the current density through the Si bar if we apply 10 V across it.

Solution

a) The concentration of electrons, $n = N_d = 10^{16}$ cm^{-3} and the concentration of holes, $p = \frac{n_i^2}{n} = \frac{(1.5\times10^{10})^2}{10^{16}} = 2.25 \times 10^4$ cm^{-3}.

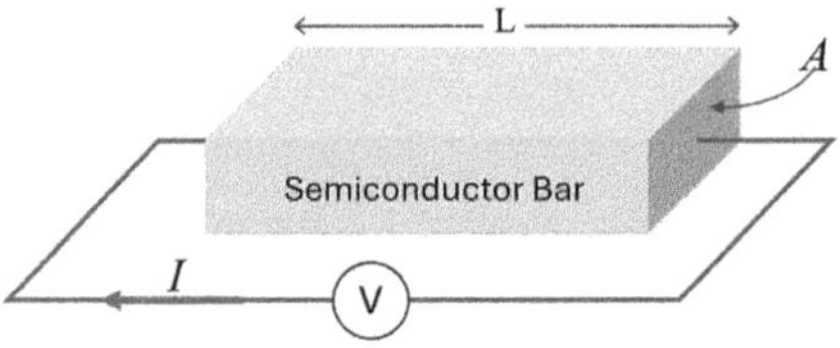

Figure 3.14 Illustration of a semiconductor bar connected to a battery.

b) The conductivity:
$\sigma = (n\mu_n + p\mu_p)\, q = (1400 \times 10^{16} + 500 \times 2.25 \times 10^{4})\, 1.6 \times 10^{-19} = 2.24\ (\Omega\ \text{cm})^{-1}$.

c) $J_{\text{total}} = (n\mu_n + p\mu_p)qE = \sigma E = \sigma \frac{V}{d} = 2.24\ (\Omega\ \text{cm})^{-1}\ \frac{10\ (\text{V})}{2\ (\text{cm})} = 11.2\ \frac{\text{A}}{\text{cm}^2}$.

Example 3.4 ***Conductivity and drift***

A n-type piece of silicon with a length of 10 cm and a cross section area of 0.05 cm × 0.05 cm is connected to 2 V potential difference. If the electron doping level is $n = 10^{17}\ \text{cm}^{-3}$ and assuming $\mu_n = 1200\ \text{cm}^2/\text{V} \cdot \text{s}$, $\mu_p = 500\ \text{cm}^2/\text{V} \cdot \text{s}$, $T = 300$ K, $n_i = 1.5 \times 10^{10}\ \text{cm}^{-3}$ and $E_g = 1.12$ eV, calculate:

a) Concentration of holes in this piece of silicon.
b) Conductivity and resistivity of the silicon.
c) The total current flowing through the silicon bar at $T = 300$ K.
d) Explain what would happen to current if temperature is doubled ($T = 600$ K) (no math)

Solution

a) Concentration of holes, $p = \frac{n_i^2}{n} = \frac{(1.5 \times 10^{10})^2}{10^{17}} = 2.25 \times 10^3\ \text{cm}^{-3}$

b) The conductivity

$$\sigma = (n\mu_n + p\mu_p)q = (1200 \times 10^{17} + 500 \times 2.25 \times 10^3) 1.6 \times 10^{-19}$$
$$= 19.2\ (\Omega\ \text{cm})^{-1}$$

Resistivity $\rho = \frac{1}{\sigma} = \frac{1}{19.2}\ (\Omega\ \text{cm}) \cong 0.052 (\Omega\ \text{cm})$

c) Current passing through the bar can be calculated from $I = A\ J_{drift}$, where $J_{drift} = J_{total}$, given as

$$J_{total} = (n\mu_n + p\mu_p)qE = \sigma E$$

Calculating the electric field across the bar with 2 V applied

$$E = \frac{2\ \text{V}}{10\ \text{cm}} = 0.2\ \text{V/cm}$$

$$J_{total} = 19.2 \times 0.2 = 3.84\ \text{A/cm}^2$$

Calculating the Area of the square cross section,

$$\text{A} = 0.05 \times 0.05 = 0.0025\ \text{cm}^2$$

$$I = 3.84 \times 0.0025 = 9.6\ \text{mA}$$

d) Increasing temperature would increase intrinsic concentration, n_i. If carrier mobility does not change significantly with temperature, then $p = n_i^2/n$ would increase with n_i. Consequently, then conductivity would increase, resulting in increased current.

3.5.5 Mobility vs. Temperature: Scattering

To understand the effect of temperature and doping on mobility, the physics of scattering mechanisms is important. There are two main scattering mechanisms in a semiconductor: lattice and impurity scattering. *Lattice scattering* is due to collective crystal vibrations from thermal energy. These vibrations are also called phonons, and hence phonon scattering can be used. When an electron or hole is drifting through a crystal lattice, vibrations can cause scattering, reducing the mobility. As the temperature increases, more thermal energy is absorbed, and the lattice scatters more frequently, reducing the mobility further. If the temperature is low, on the other hand, lattice vibrations are minimal, then phonon scattering is not significant. In this case, *ionized impurity scattering* will dominate. As the charge carrier is drifting in lattice, there is a probability it scatters with the ionized impurity. The charged electrons and holes are electrostatically attracted to the fixed charge: electrons to N_d^+ atom and holes to N_a. At low temperatures, since moving more slowly, electrons or holes have a higher probability of being in the vicinity of impurity and a higher chance of being scattered. If the temperature is high, the charge particle moves faster and will spend less time in the vicinity of the ionized impurity, and the probability of scattering is reduced. As a result, the mobility will drop with temperature, opposite to the lattice scattering trend. The trend of mobility vs temperature for lattice scattering is proportional to $T^{-3/2}$ when lattice scattering dominates and $T^{3/2}$ when impurity scattering dominates. The various mobility components can add up inversely. So, if we define μ_L as mobility to do lattice scattering, and μ_I mobility due to ionized impurity scattering, then the total mobility μ_{total} can be determined by the inverse sum of all scattering mechanisms.

$$\frac{1}{\mu_{\text{total}}} = \frac{1}{\mu_L} + \frac{1}{\mu_I} + \dots \tag{3.51}$$

Figure 3.15 plots mobility vs. temperature for lattice and phonon scattering.

At a fixed temperature, the effect of doping on mobility can be significant due to an increase in the ionized impurity scattering. Figure 3.16 plots electron and hole mobility vs. doping concentration for Si 300 K. The higher the doping the reduction in mobility due to increase in ionized impurity scattering. Also, notice the difference between electron and hole mobilities of Si since the effective mass of holes is always larger than that of electrons.

3.5.6 Effect of Electric Field on the Band Diagram

Also, it is important and interesting to see how the electric field will affect the band diagram in Figure 3.1. Remember, the band diagram represents potential energy

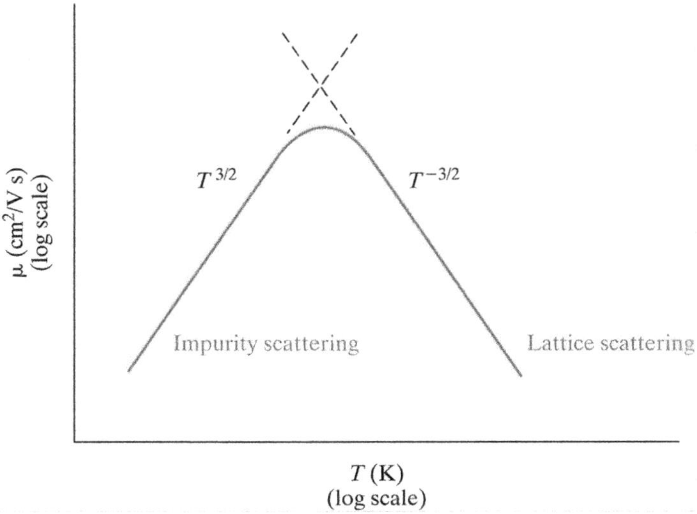

Figure 3.15 Mobility vs. temperature for impurity and lattice scattering. *Source:* [1] / with permission of Pearson Education.

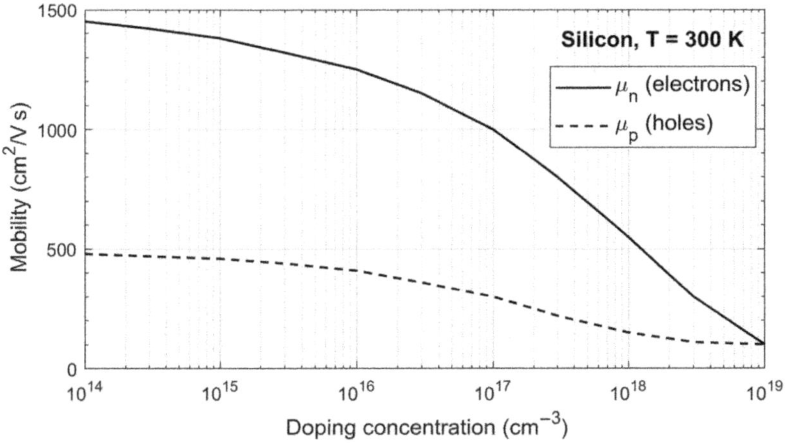

Figure 3.16 Electron and hole mobility vs. impurity concentration in silicon at T = 300K.

and the electric field is equal to the rate of change of potential energy.

$$E = -\frac{dV}{dX} \tag{3.52}$$

This implies the potential band will bend in a linear way.

Figure 3.17 shows the band diagram under electric field pointing in the negative x direction.

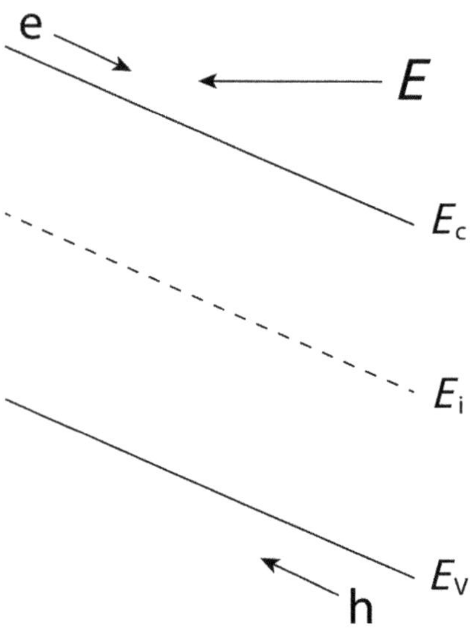

Figure 3.17 Effect of electric field on the band diagram showing band bending.

Two tips to remember the direction of the band bending. First, the electric field points into the band slope. Second, electrons drift downhill with potential. These two checks can be used to confirm correct band bending direction.

3.6 Excess Carriers

3.6.1 Optical Absorption

Up until now, we have analyzed the semiconductor under thermal equilibrium conditions with $n = p = n_i$. However, most devices operate with excess carriers above thermal equilibrium. There are two ways to get excess carriers: Optical absorption and electrical injection. It turns out, excess carriers will dominate the properties of the device, so it is critical to understand how generation mechanism of excess carriers takes place and what happens after the source of the excess carriers is removed.

Let us focus the discussion on excess carriers as it relates to optical absorption. The photoelectric effect is the emission of electrons when a photon is absorbed in a material. Electrons emitted in this manner are called photoelectrons. As a historical aside, this discovery that light can be described as photons with discrete energy packets changed the course of physics. Einstein was awarded the 1921 Nobel Prize in Physics for "his discovery of the law of the photoelectric effect," and Millikan was awarded the Nobel Prize in 1923 for "his work on the elementary charge of electricity and on the photoelectric effect." At the time, the experiment

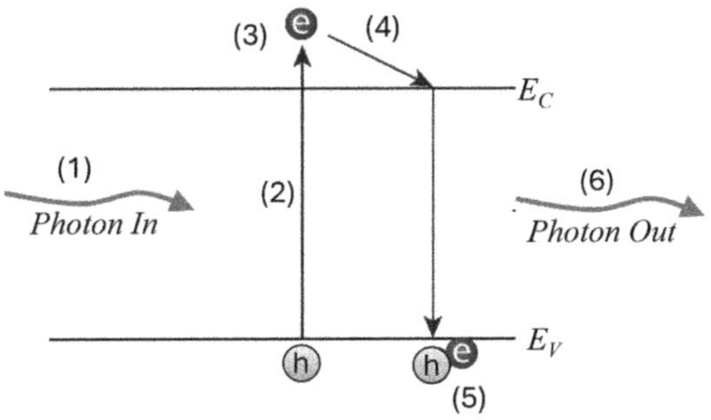

Figure 3.18 Illustration of band diagram with photon absorption.

proved that the photons can be treated as particles with certain energy and as waves (wave–particle duality). So, when a photon is absorbed in a semiconductor, one of two things may happen. If energy of the photon is equal to or larger than the bandgap, then the photon is fully absorbed in the semiconductor. As a result of that energy being absorbed, a covalent bond can break, exciting an electron from valence band to the conduction band, making an EHP. This is the same process of EHP formation due to thermal energy. These carriers that are formed from the photon absorption are called excess carriers. If the energy of the photon is less than the bandgap, then the photon will travel through the semiconductor crystal and not be absorbed (transparent). Figure 3.18 shows an energy band diagram with an incoming photon and the various processes that happen after.

When the electron recombines directly from the conduction band to valence band, the photon is emitted, and this process is called photoluminescence.

Below are the steps leading to photoluminescence as shown in Figure 3.18.

Step 1: Photon with energy more than the bandgap is absorbed.
Step 2: EHP is created.
Step 3: Electron makes it to conduction band.
Step 4: Electron gives up energy in the lattice to scattering events.
Step 5: Electron returns to valence band and recombines with hole.
Step 6: A photon is released with energy of the bandgap.

3.6.2 Indirect and Direct Bandgap: Traps

Going back to formation of the bandgap in Section 3.1, it turns out that for semiconductors, there are two distinct classes: indirect and direct bandgap.

In a direct bandgap semiconductor, an electron that has been excited into the conduction band can return to the valence band in a single step. So, the electron–hole recombination is very fast. When this happens, the excess energy is released directly as a photon. Since this transition is simple and fast, light emission is very efficient. Materials such as GaAs belong to this category and are therefore widely used in LEDs and laser diodes.

In an indirect bandgap semiconductor, the situation is different. An electron in the conduction band cannot return directly to the valence band in one step. Instead, the recombination process requires additional intermediate interactions with the crystal lattice. As a result, the energy is often released as heat rather than light, making photon emission much less likely. Silicon is the most common example of an indirect bandgap semiconductor, which explains why it is a poor light emitter despite being excellent for electronic devices.

Up until now, we treated the bandgap as forbidden, and no electrons are allowed to reside within. But it turns out impurities can reside in the bandgap. These impurity levels during the fabrication process can arise from contamination or crystal defects. These levels reside in the bandgap of the semiconductor. An electron in the conduction band can recombine with the hole via a trap state. Figure 3.19 shows the band diagram with the trap state and the process of photon absorption.

Steps of photon absorption when a trap state is present:

Step 1: Photon with energy more than the bandgap is absorbed.
Step 2: EHP is created.
Step 3: Electron makes it to conduction band.
Step 4: Electron gives up energy in the lattice to scattering events.
Step 5: Electron can now go from conduction to trap.
Step 6: Electron returns to conduction band.
Step 7: Electron now returns to valence band.
Step 8: Photon is released.

It is clear the trap level will increase the time it takes for the electron to return to valence band. This delays the recombination and reduces the photon energy that is emitted.

3.6.3 Lifetime

What happens to the excess carriers after they are created? The electron will eventually return to the valence band and recombine with the hole. The excess

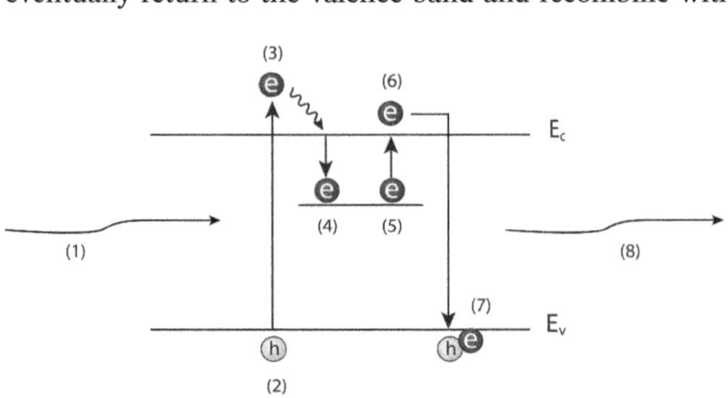

Figure 3.19 Illustration of photon absorption with trap state (see steps 1–8).

electrons that are in the conduction band can recombine with holes either through a trap state indirectly or directly from without the trap state. In the direct process, the energy lost by the electron in making the transition is given up as a photon. This happens spontaneously, meaning the probability that electron will recombine with the hole is constant. When the semiconductor absorbs light, the additional EHPs are created beyond thermal equilibrium. The physical process of this decay will be *exponential* with time by

$$\Delta n(t) = \Delta n(0)e^{-t/\tau_n} \tag{3.53}$$

where $\Delta n(0)$ is the initial concentration at $t = 0$, before the excess carriers are created. τ_n is one of the most important properties of semiconductors called the *electron minority carrier lifetime.* It physically means how long the minority carrier electron will last before it recombines with the hole. For excess holes,

$$\Delta p(t) = \Delta p(0)e^{-t/\tau_p} \tag{3.54}$$

where τ_p is the *hole minority carrier lifetime.* The time signifies how long the minority carrier holes will last before it recombines with the electron. At a steady-state optical excitation, excess carriers can be calculated based on the carrier lifetime and optical generation rate, g_{op}, as follows:

$$\delta n = \tau_n \cdot g_{op} \text{ and } \delta p = \tau_p \cdot g_{op} \tag{3.55}$$

So total steady-state carrier concentration can be found as

$$n = n_0 + \delta n \text{ and } p = p_0 + \delta p \tag{3.56}$$

Example 3.5 ***Lifetime and excess carriers***
Let us consider a sample of Ge that is doped with 10^{17} cm^{-3} acceptors that is briefly exposed to incident light that results in a generation of 10^{14} EHP/cm^3 (at time $t = 0$). If minority carrier lifetimes $\tau_p = \tau_n = 1$ μs, find:

a) Total concentration of holes and electrons at time $t = 0$.
b) total concentration of holes and electrons 2 μs after light source is removed.

Solution

a) Intrinsic concentration in Ge is $n_i = 2.5 \times 10^{13}$ cm^{-3}. Since Ge is doped with acceptors, this is p-type material and $N_a = p_0 = 10^{17}$ cm^{-3}.
It follows that $n_0 = \frac{n_i^2}{p_0} = \frac{(2.5\times10^{13})^2}{10^{17}} = 6.25 \times 10^9$ cm^{-3}. p_0 and n_0 are carrier concentrations in Ge sample prior to exposure to the light source.
At time $t = 0$, excess carriers (EHP) are generated: $\Delta p(0) = \Delta n(0) = 10^{14}$ cm^{-3}.
So total concentration of each of the carriers at $t = 0$ is

$$p(0) = p_0 + \Delta p(0) = 10^{17} + 10^{14}\,\text{cm}^{-3} \approx 10^{17}\,\text{cm}^{-3}$$

$$n(0) = n_0 + \Delta n(0) = 6.25 \times 10^9 + 10^{14}\,\text{cm}^{-3} \approx 10^{14}\,\text{cm}^{-3}$$

b) After time of $t = 2\ \mu s$ have passed, excess carriers will decay as in Eq. (3.54)

$$p(t) = p_0 + \Delta p(t) = p_0 + \Delta p(0)e^{-t/\tau_p} = 10^{17} + 10^{14}e^{-\frac{2\times10^{-6}}{10^{-6}}}\ \text{cm}^{-3}$$
$$p(t) = 10^{17} + 1.35 \times 10^{13}\ \text{cm}^{-3} \cong 10^{17}\ \text{cm}^{-3}$$

and

$$n(t) = n_0 + \Delta n(t) = n_0 + \Delta n(0)e^{-t/\tau_n} = 6.25 \times 10^{9} + 10^{14}e^{-\frac{2\times10^{-6}}{10^{-6}}}\ \text{cm}^{-3}$$
$$n(t) = 6.25 \times 10^{9} + 1.35 \times 10^{13}\ \text{cm}^{-3} \cong 1.35 \times 10^{13}\ \text{cm}^{-3}$$

3.6.4 Quasi-Fermi Level

When there are excess carriers, the normal Fermi level will not capture the effect on the excess minority carriers. A mathematical way to track the concentration is by using quasi-Fermi levels, E_{fn} for electrons and E_{fp} for holes defined as for electrons,

$$n = n_i e^{\frac{(E_{fn} - E_i)}{kT}} \tag{3.57}$$

and for holes,

$$p = n_i e^{\frac{(E_i - E_{fp})}{kT}} \tag{3.58}$$

Figure 3.20 shows an example of the quasi-Fermi levels for both electrons and holes.

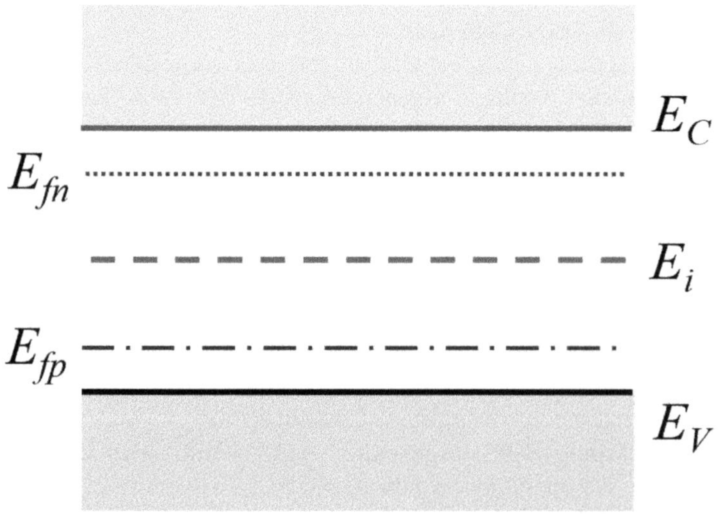

Figure 3.20 Band diagram showing the quasi-Fermi levels for both electrons and holes.

Example 3.6 *Quasi-fermi levels*
Given n-type Si sample with donor doping of $N_d = 10^{16}$ cm^{-3}. Sample is exposed to a light source that results in the generation of 10^{19} cm^{-3} EHP/s uniformly across the sample. Assume that electron and hole lifetimes are both equal to 1 μs, diffusivities $D_p = 12$ cm^2/s, and $D_n = 36$ cm^2/s. Find the quasi-Fermi levels for this sample.

Solution
At room temperature, $n_i \approx 10^{10}$ cm^{-3} and $n_0 = N_d = 10^{16}$ cm^{-3}. At thermal equilibrium, we have that

$$p_0 = \frac{n_i^2}{n_0} = \frac{(10^{10}\ \text{cm}^{-3})^2}{10^{16}\ \text{cm}^{-3}} = 10^4\ \text{cm}^{-3}$$

At a steady-state optical excitation with $g_{op} = 10^{19}$ EHP/cm^3 s and carrier lifetimes $\tau_n = \tau_p = 1$ μs, excess electron and hole carriers can be calculated as

$$\delta n = \delta p = \tau \cdot g_{op} = 1 \times 10^{-6}\ \text{s} \times 10^{19}\ \text{EHP/cm}^3\text{s} = 10^{13}\ \text{cm}^{-3}$$

To find total carrier concentration, we have $n = n_0 + \delta n$ and $p = p_0 + \delta p$. So,

$$n = n_0 + \delta n = 10^{16}\ \text{cm}^{-3} + 10^{13}\ \text{cm}^{-3} = 1.001 \times 10^{16}\ \text{cm}^{-3}$$
$$p = p_0 + \delta p = 10^4 + 10^{13}\ \text{cm}^{-3} = 10^{13}\ \text{cm}^{-3}$$

Total carrier concentration in terms of quasi-Fermi levels is given by

$$n = n_i e^{(E_{fn} - E_i)/kT} \Rightarrow E_{fn} - E_i = kT \cdot \ln\left(\frac{n}{n_i}\right)$$
$$p = n_i e^{(E_i - E_{fp})/kT} \Rightarrow E_i - E_{fp} = kT \cdot \ln\left(\frac{p}{n_i}\right)$$

where $kT = 0.0259$ eV for room temperature.
Then,

$$E_{fn} - E_i = kT \ln\left(\frac{n}{n_i}\right) = 0.0259 \ln\left(\frac{1.001 \times 10^{16}}{10^{10}}\right) = 0.358\ \text{eV}$$
$$E_i - E_{fp} = kT \ln\left(\frac{p}{n_i}\right) = 0.0259 \ln\left(\frac{10^{13}}{10^{10}}\right) = 0.179\ \text{eV}$$

3.7 Diffusion

3.7.1 Diffusion Current

The excess carriers created will form concentration gradient with a region of very high concentration and regions of lower concentration. This gradient starts a diffusion of carriers. The diffusion process happens in everyday life such as an odor in a

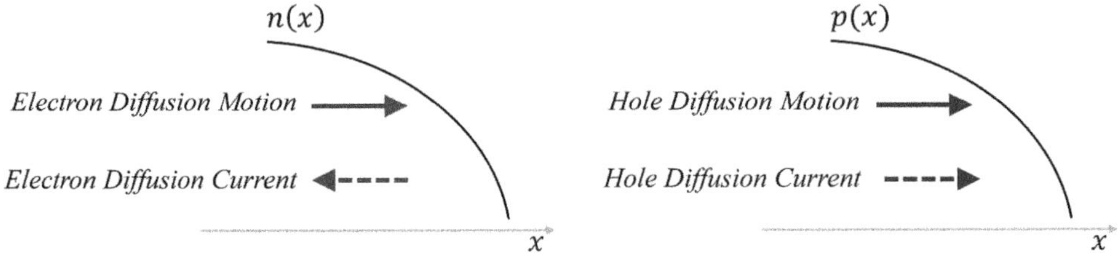

Figure 3.21 Direction of diffusion flux and diffusion current due to a concentration gradient.

room that starts at one side. The odor will move from the source and spread around the room. Eventually, the room is covered uniformly with odor and non-odor molecules. We can say the odor molecules *diffused* from high concentration to areas of low concentration. This analogy can be used to describe a semiconductor with excess carriers. Carriers will move or diffuse from regions of high concentration of carriers to regions of low concentration until there is a uniform number of carriers. This motion of the carriers results in a diffusion current. The mathematical derivation for diffusion current is done by modeling it as a diffusion flux density over space. We can define $\bar{l}$ as the mean free path between collisions and $\bar{\tau}$ as time between collisions during the diffusion of the charged carriers. The diffusion coefficient is defined for both holes and electrons with units cm^2/s

$$D_{n,p} \equiv \frac{\bar{l}^2}{\bar{\tau}} \tag{3.59}$$

For electron diffusion, the concentration gradient multiplied by diffusion coefficient and the charge of one electron (q) is the diffusion current density for electrons.

$$J_n = qD_n \frac{dn(x)}{dx} \tag{3.60}$$

We have the same derivation for hole diffusion current.

$$J_p = -qD_p \frac{dp(x)}{dx} \tag{3.61}$$

Note that since electrons and holes move in the same direction, the *diffusion current* will be in opposite directions due to negatively charged electrons and positively charged holes. Figure 3.21 shows the direction of diffusion flux and diffusion current due to a concentration gradient.

3.7.2 Recombination and Diffusion Length

As the charge carriers diffuse, recombination takes place. In a steady state, the diffusion of the excess carriers will decay exponentially with distance like drift with lifetime. For holes diffusion with $\Delta p(0)$ as the initial concentration $x = 0$, then

$$\Delta p(x) = \Delta p(0)e^{-x/L_p} \tag{3.62}$$

L_p is the diffusion length for holes, given by units of cm, and related to diffusion coefficient and lifetime by

$$L_p = \sqrt{D_p \tau_p} \tag{3.63}$$

The physical meaning of diffusion length is the distance the hole moves before it recombines. For electron diffusion with $\Delta n(0)$, as the initial concentration $x = 0$, then

$$\Delta n(x) = \Delta n(0)e^{-x/L_n} \tag{3.64}$$

L_n is the diffusion length for electrons given by

$$L_n = \sqrt{D_n \tau_n} \tag{3.65}$$

The diffusion length represents the average distance an excess electron or hole diffuses before recombining.

Example 3.7 ***Diffusion current***

Consider a bar of silicon in which a hole concentration profile described by

$$p(x) = p_0 e^{-x/L_p}$$

where x is the length across the bar, $p_0 = 10^{17}$ cm^{-3}, $L_p = 0.1$ μm, and $D_p = 12$ cm^2/s.

Find the diffusion hole-current density, J_p at $x = 0$ and at $x = 1$ μm.

Solution

$$J_p = -qD_p \frac{dp(x)}{dx} = -qD_p \frac{d}{dx}\left(p_0 e^{-\frac{x}{L_p}}\right)$$

$$= -qD_p p_0 e^{-\frac{x}{L_p}}\left(-\frac{1}{L_p}\right) = \frac{qD_p p_0}{L_p} e^{-\frac{x}{L_p}}$$

Substituting the values and making sure that all units are in cm^2 we get.

$$L_p = 0.1\ \mu\text{m} = 0.1 \times 10^{-4}\ \text{cm} = 10^{-5}\ \text{cm}$$

$$J_p(x = 0) = \frac{qD_p p_0}{L_p} = \frac{1.6 \times 10^{-19} \times 12 \times 10^{17}}{10^{-5}}\left(\frac{\text{A}}{\text{cm}^2}\right)$$

$$= 1.92 \times 10^4\ \text{A/cm}^2$$

$$J_p(x = 1) = \frac{qD_p p_0}{L_p} e^{-\frac{x}{L_p}} = \frac{1.6 \times 10^{-19} \times 12 \times 10^{17}}{10^{-5}} e^{-\frac{1}{0.1}}\left(\frac{\text{A}}{\text{cm}^2}\right)$$

$$= 0.872\ \text{A/cm}^2$$

3.7.3 Total Current (Drift and Diffusion)

If a semiconductor is exposed to an electric field and concentration gradient, this total current can be a combination of both diffusion current and drift current for both electrons and holes. We can calculate both the electron and hole current component separately and add it up to get the total current. Mathematically speaking, the total current is equal to the hole component plus the electron component. The electrons and holes will both have drift and diffusion component. However, the direction of the current due to the component's needs to be looked at from a physics perspective, depending on the electric field direction and concentration gradient. We can capture this by using a mathematical sign of positive or negative. So, let us write the total formula now and get one final equation for the total current.

$$J_{total} = J_n + J_p \tag{3.66}$$

Each carrier has both drift and diffusion.

$$J_{n/p} = J_{drift} + J_{diffusion} \tag{3.67}$$

The total electron current is

$$J_n = q\mu_n n(x)\mathrm{E}_{\mathrm{x}} + qD_n \frac{dn(x)}{dx} \tag{3.68}$$

And the total hole current is

$$J_p = q\mu_p p(x)\mathrm{E}_{\mathrm{x}} - qD_p \frac{dp(x)}{dx} \tag{3.69}$$

Example 3.8 ***Total current (drift + diffusion due to electrons)***

Consider the bar of silicon described in Example 3.7 is subjected to constant electric field of 0.2 V/cm. Calculate total current density at $x = 1$ μm, assuming hole mobility $\mu_p = 500\ \mathrm{cm^2/V \cdot s}$

Solution

Note that

$$p(x = 1\ \mu\mathrm{m}) = p_0 e^{-x/L_p} = 10^{17} e^{-10}\ \mathrm{cm^{-3}} = 4.54 \times 10^{12}\ \mathrm{cm^{-3}}$$

Using (Eq. (3.69))

$$J_{p,diffusion}(x = 1) = -qD_p \frac{dp(x)}{dx} = \frac{qD_p p_0}{L_p} e^{-\frac{x}{L_p}} = 0.872\ \mathrm{A/cm^2}$$

$$J_{p,total} = 1.6 \times 10^{-19} \times 500 \times 4.54 \times 10^{12} \times 0.2 + 0.872\ \left(\frac{\mathrm{A}}{\mathrm{cm^2}}\right)$$

$$J_{p,total} = 7.26 \times 10^{-5} + 0.872\ \left(\frac{\mathrm{A}}{\mathrm{cm^2}}\right) \cong 0.872\ \left(\frac{\mathrm{A}}{\mathrm{cm^2}}\right)$$

Notice that in this case total current is diffusion dominated. For given doping profile, larger electric field is required to have a larger contribution from the drift current.

3.7.4 Einstein Relationship

As one would expect from the physics of diffusion and drift process, diffusivity and mobility are related via the Einstein relationship which states

$$\frac{D}{\mu} = \frac{kT}{q} \tag{3.70}$$

This formula (Eq. (3.70)) is easy to remember because of the mnemonic rhyme "Dee over Mu, Is Kay Tee over Q" (pronounced dee over myoo, is kay tee over kyoo).

Example 3.9 ***Einstein relationship***
If the mobility of holes is equal to 1500 $cm^2/V\,s$ at 300 K for a given semiconductor, find the value of the hole diffusivity D_p.

Solution

$$\frac{D}{\mu} = \frac{kT}{q}, \quad D_p = \mu_p \frac{kT}{q} = 1500 \times 0.0259 = 38.85\ \text{cm}^2/\text{s}$$

Example 3.10 ***Diffusion length***
A silicon solar cell has a doping concentration of 10^{16} cm^{-3} for both the p-type and n-type regions. The carrier lifetime for both, electrons and holes, is τ =1 μs. Calculate the diffusion length, L_n and L_p, for the minority carriers (electrons and holes) in the silicon solar cell. Assume diffusivity of holes and electrons to be equal to 30 cm^2/s.

Solution
Given:

$$N_a = N_d \approx n = p = 10^{16}\ \text{cm}^{-3}$$

$$\tau_n = \tau_p = 10^{-6}\ \text{s}$$

We have that $L_n = \sqrt{D_n \tau_n}$ and since doping, lifetime, and diffusivity are all equal.

Then,

$$L_n = L_p$$

$$L_p = L_n = \sqrt{D_n \tau_n} = \sqrt{30\ (\text{cm}^2/\text{s}) \times 10^{-6}\ \text{s}} = 0.0055\ \text{cm}$$

3.8 Chapter Summary

1) Semiconductors are the material building blocks of solar cells and are used to convert optical energy from photons into electrical energy. Understanding semiconductor physics is essential to understanding how solar cells operate.
2) During crystal formation, the interaction of atomic electron wavefunctions leads to the formation of energy bands, consisting of a valence band, a conduction band, and a forbidden energy bandgap. The size of the bandgap determines the electrical and optical properties of a material.
3) Based on the size of the bandgap, materials are classified as metals, insulators, or semiconductors. Metals have overlapping bands, insulators have large bandgaps, and semiconductors have moderate bandgaps that allow controlled conduction.
4) When sufficient energy is absorbed, electrons can transition from the valence band to the conduction band, leaving behind an empty band that is treated as a positively charged hole. Each excitation creates an EHP.
5) In thermal equilibrium, the rate of generation of EHPs equals the rate of recombination, resulting in an intrinsic carrier concentration that depends on the bandgap and temperature of the material.
6) The electrical properties of semiconductors can be modified by doping, where donor atoms add electrons and acceptor atoms add holes. Doped materials are called extrinsic semiconductors and can be classified as n-type or p-type depending on the dominant carrier.
7) Carrier concentrations in the conduction and valence bands are determined using Fermi–Dirac statistics. The Fermi level provides a convenient way to describe the probability of electron occupancy and shifts with doping and temperature.
8) Charge carriers move under the influence of an electric field through a process called drift, producing a drift current. Carrier mobility quantifies how easily electrons or holes move through the crystal and is influenced by scattering mechanisms.
9) In the presence of concentration gradients, carriers diffuse from regions of high concentration to regions of low concentration, producing a diffusion current. The diffusion coefficient and mobility are related through the Einstein relation.
10) Excess carriers are generated by optical absorption or electrical injection and dominate device behavior under operation. These carriers recombine over time with a characteristic minority carrier lifetime.
11) Under nonequilibrium conditions, separate quasi-Fermi levels are used to describe electron and hole populations. These concepts are essential for analyzing semiconductor devices such as diodes and solar cells.

End-of-Chapter Problems

Short Questions

P3.1 A semiconductor where holes are the majority carriers is a *p*-type semiconductor?

P3.2 A semiconductor where electrons are the majority carriers is a *p*-type semiconductor?

P3.3 The valence electron of a conductor can also be called a _____.

P3.4 When silicon atoms combine to form a solid, they arrange themselves into an orderly pattern called_________.

P3.5 A doped semiconductor is called an intrinsic semiconductor.
True/False

P3.6 Doping is a process of adding impurity atoms to an intrinsic crystal to alter its electrical conductivity.
True/False

P3.7 Intrinsic semiconductor has unequal number of free electrons and holes.
True/False

P3.8 A pure semiconductor is also referred to as____________.

P3.9 A doped semiconductor is also referred to as____________.

P3.10 Recombination is when____________.

P3.11 At 0 K temperature, semiconductor behaves ______________.

Numerical Problems

P3.12 Calculate the hole concentration in an *n*-type Si with concentration of donors equal to 10^{15} cm^{-3}? (Assume room temperature).

P3.13 Calculate the intrinsic carrier concentration of Germanium at $T = 300$ K and $T = 600$ K. Assume that for Ge, $B = 1.66 \times 10^{15}$ and $E_g = 0.68$ eV are constant with temperature.

P3.14 Given an undoped Si material with bandgap $E_g = 1.12\,\text{eV}$, material constant $B = 7.3 \times 10^{15}\ \text{cm}^{-3}\ \text{K}^{-\frac{3}{2}}$, and Boltzmann's constant $k = 8.62 \times \frac{10^{-5}\ \text{eV}}{\text{K}}$.

a) Calculate intrinsic carrier concentration of Si at $T = 600\ \text{K}$.

b) What is the concentration of electrons and holes in intrinsic Si?

P3.15 A n-type piece of silicon with a length of 10 cm and a cross-sectional area of 0.05 cm × 0.05 cm is connected to 2 V potential difference. If the electron doping level is $n = 10^{17}\ \text{cm}^{-3}$, calculate the total current flowing through the silicon bar at $T = 300\ \text{K}$. (Assume $\mu_n = 1200\ \text{cm}^2/\text{V} \cdot \text{s}$, $\mu_p = 500\ \text{cm}^2/\text{V} \cdot \text{s}$, $T = 300\ \text{K}$, $n_i = 1.5 \times 10^{10}\ \text{cm}^{-3}$, and $E_g = 1.12\ \text{eV}$.)

P3.16 For a Si bar doped p-type at $10^{18}\ \text{cm}^{-3}$, and at a temperature of $T = 400\ \text{K}$, calculate:

a) Intrinsic concentration

b) Concentration of electrons

c) Conductivity of the bar

Assume that the electron and hole mobilities at 400 K are $1600\ \text{cm}^2/\text{V} \cdot \text{s}$ and $600\ \text{cm}^2/\text{V} \cdot \text{s}$, respectively.

P3.17 Assume room temperature. Draw and label (E_g, E_f, E_i, E_c, and E_v) the energy band diagram of Ge if it is doped with $10^{17}\ \text{cm}^{-3}$ acceptors.

P3.18 A Si sample is doped with 10^{16} donor atoms/cm^3. Find the hole concentration at equilibrium, p_0, at 400 K? Where is E_F relative to E_i? Sketch the band diagram for this material.

P3.19 A Si sample is doped with $10^{16}\ \text{cm}^{-3}$ Boron and a certain number of donor atoms (N_d). The Fermi level is 0.36 eV above E_i at 300 K. Calculate value of N_d.

P3.20 Light is shone on Ge sample that is doped with acceptors of concentration $10^{17}\ \text{cm}^{-3}$. This results in uniform optical generation of 10^{20} EHPs/cm^3 at $t = 0$. Assume hole and electron lifetime $\tau_p = \tau_n = 1\ \mu\text{s}$. Calculate Fermi level before shining light and quasi-Fermi levels after exposure to light.

P3.21 Consider a silicon solar cell where the concentration gradient of holes in the p-region is given by $dp/dx = -10^{21}\ \text{cm}^{-4}$. The diffusion coefficient for holes in silicon, $D_p = 12\ \text{cm}^2/\text{s}$. Calculate the hole diffusion current density J_p^{diff}.

Reference

1 Streetman, B.G. and Banerjee, S. (2015). Solid State Electronic Devices, Globale. Prentice Hall.

Reading Materials

1 Neamen, D. (2011). Semiconductor Physics and Devices: Basic Principles, 4th ed. McGraw-Hill.
2 Sze, S.M., Li, Y., and Ng, K.K. (2021). Physics of Semiconductor Devices, 4th ed. Wiley.

4

Physics of the PN Junction

4.1 About the Chapter

To make semiconductor-based devices, *p*- and *n*-type layers need to be combined to form junctions. The *pn* junction diode is the fundamental building block of many important semiconductor devices. A solar cell is essentially a *pn* junction designed to absorb photons coming from the sun and convert them to electrons. As a result, it is vital to study the physics of the *pn* junction to understand how a solar cell works. In this chapter, the *pn* junction physics is explained starting from the built-in potential to the electrostatics, and the effect of applied voltage. Namely, the components of electrical current are studied due to the motion of electrons and holes after applying a forward and reverse bias voltage. This leads to the ideal *pn* junction diode current–voltage characteristics. Finally, some real effects are discussed and how they affect the *pn* diode ideal current–voltage characteristics.

Key Topics

- *pn* junction formation and built-in potential
- Depletion region and junction electrostatics
- Energy band diagram of the *pn* junction
- Forward and reverse bias operation
- Ideal and nonideal diode characteristics

4.2 *pn* Junction

4.2.1 Equilibrium and Built-In Potential

Let us start with a simple thought experiment using the physics from the previous chapters. Imagine a *p*-type semiconductor and an *n*-type semiconductor close to

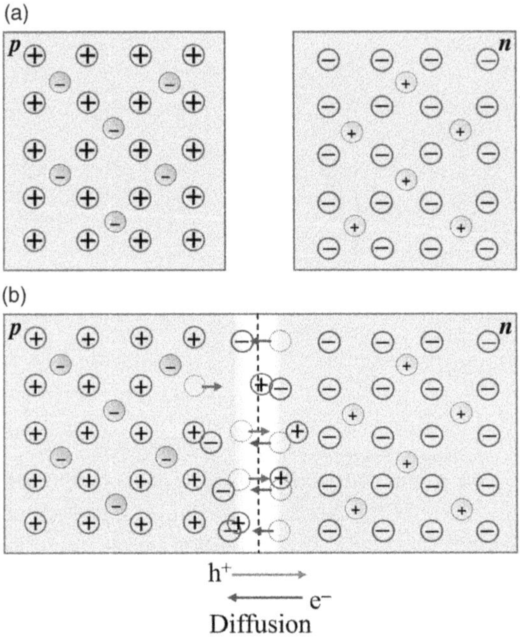

Figure 4.1 *pn* junction thought experiment with (a) *p*- and *n*-sides shown separately and (b) *p*- and *n*-sides connected, showing the diffusion process between *p*- and *n*-sides of a junction.

each other but separated. In this initial thought experiment, no transfer of charge can take place due to the air gap separating them. Now imagine the *p* and *n* layers are connected, what happens next? Remember the *p*-type semiconductor has a majority of holes with minority electrons and an *n*-type semiconductor has a majority of electrons with minority holes. This creates a concentration gradient with holes diffusing from the *p*-side to the *n*-side and the electrons diffusing from the *n*-side to the *p*-side. Figure 4.1 shows the cartoon of the thought experiment.

Eventually, after a long period of time, one would expect that the semiconductor will become a mix of electrons and holes as the diffusion continues. However, something prevents this from happening and stops the charge build up. To understand this, let us analyze the charge around the junction. When the hole diffuses from the *p*- to *n*-side, negatively charged ionized impurity atoms (N_a^-) are left behind. When the electron diffuses from the *n*- to *p*-side, positively charged ionized impurity atoms (N_d^+) are left behind. The fixed (immobile) dopant atoms lose neutrality and become either positively or negatively charged when the electron or hole leaves its vicinity. This leaves an area of negative charge on the *p*-side and an area of positive charge on the *n*-side. This charge dipole will create an electric field from *n* to *p* (from positive charge to negative charge). This is called the built-in electric field. So simply by connecting *p*- and *n*-type semiconductors, a built-in electric field is induced, which is also called the contact potential. Figure 4.2 shows the *pn* junction, with charges and depletion regions.

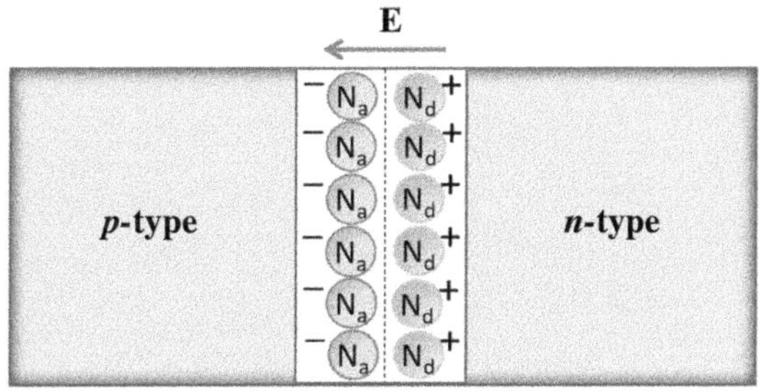

Figure 4.2 Illustration of a *pn* junction, with indicated immobile charges (ionized impurities) and the depletion region.

This built-in electric field creates a drift of electrons and holes. Since the electric field direction is from the *n*− to *p*-side electrons will drift from the *p*- to *n*-side since the negatively charged electrons move in the opposite direction to the electric field. The positively charged holes will drift from *n* to *p* side. This is in the opposite direction to the motion of diffusion due to the concentration gradient. This means the drift of carriers due to electric field has the effect of canceling out the diffusion and preventing the mix of carriers that was predicted. Figure 4.3 shows the junction with ionized impurities and the direction of electron/hole diffusion and drift.

Also, since no net current is flowing in this device, the amount of electron drift must equal the amount of electron diffusion. And the amount of hole drift must equal the amount of hole diffusion. At that point, there exists an electric field value where drift is balanced by diffusion, and at this point, the maximum number of ionized fixed doping charges is reached. We can define this point as the maximum built-in electric field or more commonly referred to as the built-in potential (V_0). The region with fixed ionized impurities is called depletion region or space charge region with thickness of W. Figure 4.4 shows final *pn* junction with labeled depletion region and built-in potential curve.

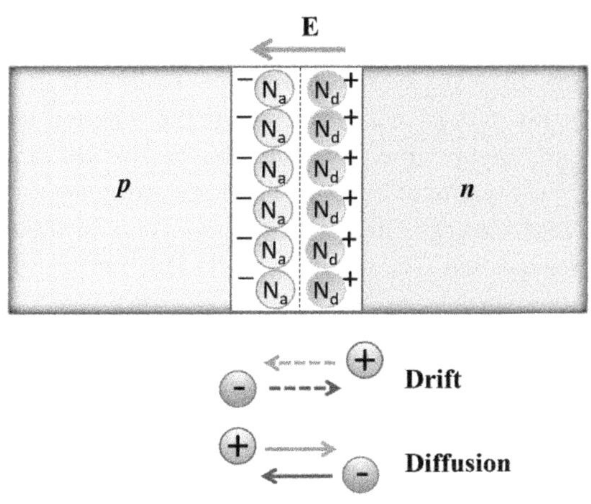

Figure 4.3 Illustration of a *pn* junction with ionized impurities and the direction of electron/hole diffusion and drift.

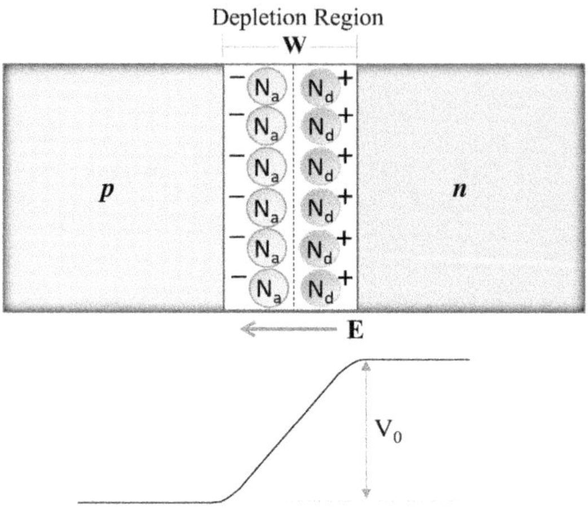

Figure 4.4 *pn* junction showing the depletion region and plotting the built-in potential.

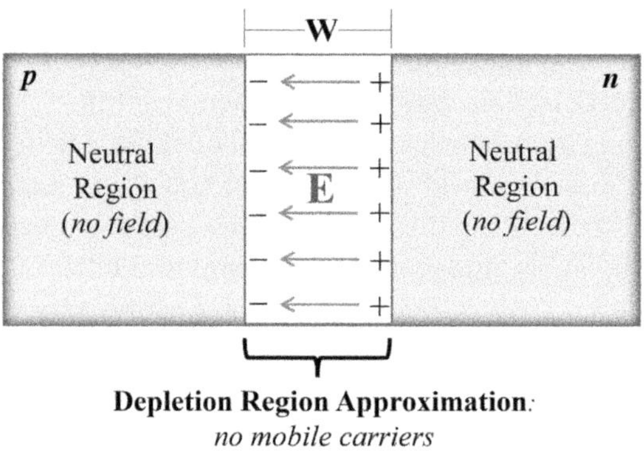

Figure 4.5 Visualization of the depletion approximation in a *pn* junction.

Let us take a closer look at the charges around the junction. On *p*-side, there are fixed ionized impurities (N_a^-) in addition to the electrons that are swept from *n- to p*-side. On *n*-side, there are fixed ionized impurities (N_d^+), in addition, the holes that are swept from *p- to n*-side. The mobile holes/electrons are swept quickly out of the junction by the electric field. One approximation is that at any given time, the only charges in the depletion region are fixed ionized impurities since the mobile charges are swept out quickly. This is called *depletion approximation*. The electric field is zero in the neutral regions and non-zero in the depletion region. Figure 4.5 is a visualization of the depletion approximation.

4.2.2 *pn* Junction Band Diagram

To understand the physics of the *pn* junction a band diagram is helpful. The first step is to draw the Fermi level as a constant across junction. In thermal equilibrium, there is no current flowing and zero net transfer of energy. So, the Fermi level is continuous across the junction to satisfy equilibrium. This means there is no slope or gradient in the Fermi level, E_f. The next step is to draw the intrinsic Fermi level E_i location relative to Fermi level on *p*-side and *n*-side away from the depletion region. These values, namely $(E_f - E_i)$ on *n*-side and $(E_i - E_f)$ on *p*-side, can be calculated from Eqs. (3.20) and (3.22), respectively. Next, we draw valence and conductions bands, E_v and E_c, on both *p*- and *n*-sides, which are placed $E_g/2$ eV away from E_i energy level. Finally, E_c, E_i, and E_v are connected from *p*- to *n*-side, forming the band diagram, which also illustrates the built-in potential at the junction. Figure 4.6 shows progression steps of drawing the *pn* junction

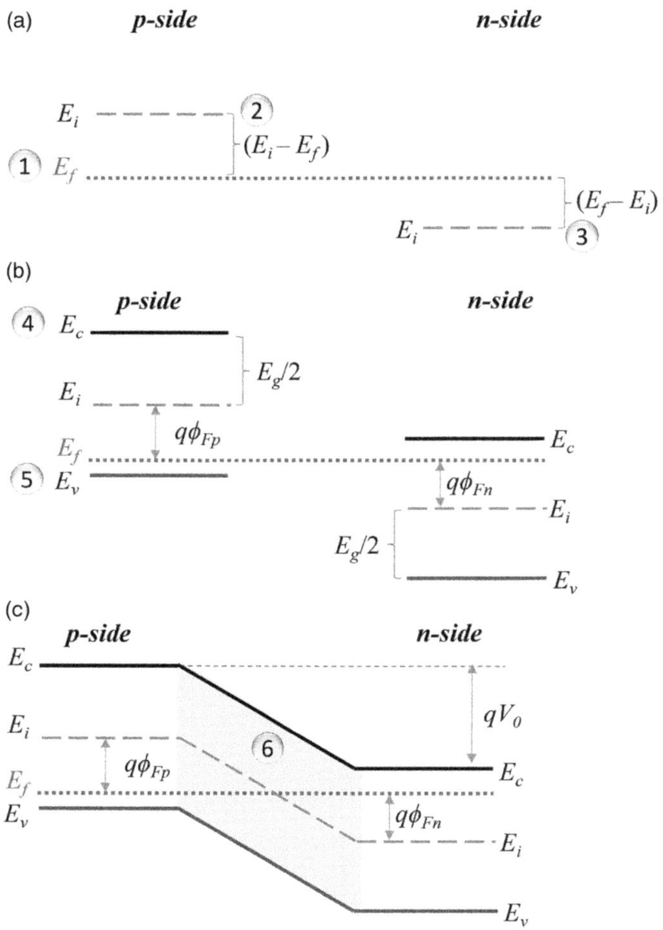

Figure 4.6 Illustrations of the steps for drawing band diagram of a *pn* junction: (a) steps 1–3, (b) steps 4 and 5, and (c) last step of connecting *p*- and *n*-sides.

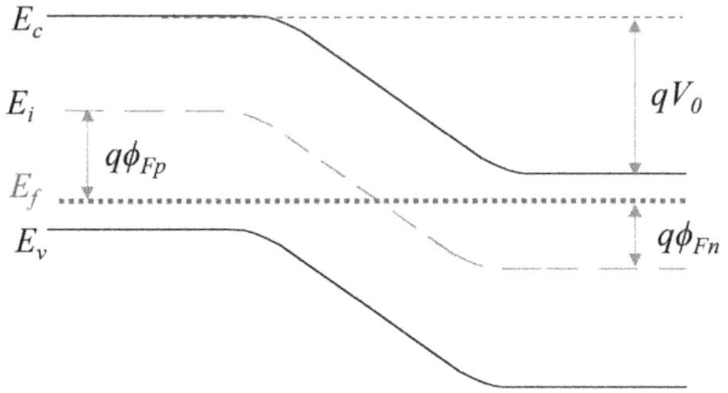

Figure 4.7 Final energy band diagram of a *pn* junction with all relevant energy levels labeled.

described above. From the band diagram, the drop in potential energy from *p*- to *n*-side is clearly shown and is due to the built-in electrical field. This potential barrier creates a diffusion barrier equal to V_0. Figure 4.7 shows final labeled *pn* junction energy band diagram.

The built-in potential can be calculated using

$$V_0 = (E_f - E_i)_{n\text{-side}} + (E_i - E_f)_{p\text{-side}} \tag{4.1}$$

This potential barrier reduces the diffusion of electrons from *n*- to *p*-side and diffusion of holes from *p*- to *n*-side. The barrier represents the built-in electric field. Electrons drift downhill since they are negatively charged, while holes drift uphill since they are positively charged on the energy band diagram. Figure 4.8 shows the motion of carriers due to both the built-in field and diffusion.

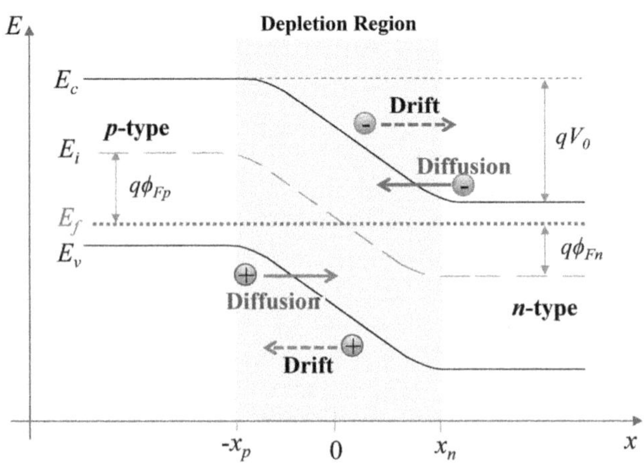

Figure 4.8 Illustration of the motion of carriers due to the built-in field, diffusion, and drift.

Example 4.1 ***Calculate and draw band diagram***
Take a Si *pn* junction doped $N_d = 1 \times 10^{18}$ cm^{-3} $N_a = 1 \times 10^{16}$ cm^{-3}

a) Calculate the separation in the Fermi level on each side of the junction
b) Draw the brand diagram
c) Extract V_0 from the diagram

Solution
Using Eq. (3.20)

$$n_0 = n_i e^{(E_f - E_i)/kT}$$

On the *n-side*, solve for $E_f - E_i$ for the electron concentration

$$\frac{n_0}{n_i} = e^{(E_f - E_i)/kT}$$

$$ln\frac{n_0}{n_i} = \frac{(E_f - E_i)}{kT}$$

$$kT \times ln\frac{n_0}{n_i} = E_f - E_i \tag{4.2}$$

Plugging in kT at room temperature is 300 K × 8.62 × 10^{-5} eV/K = 0.0259 eV.
n_i at room temperature for Si is 1.5 × 10^{10} cm^{-3}.
Since $N_d \gg n_i$ that means the doping dominates the overall number of carriers.
So, we can say $n_0 \cong N_d = 1 \times 10^{18}$ cm^{-3}.
Using Eq. (4.2),

$$E_f - E_i = kT \times ln\frac{n_0}{n_i}$$

$$E_f - E_i = 0.0259\ \text{eV} \times ln\frac{1 \times 10^{18}\ \text{cm}^{-3}}{1.5 \times 10^{10}\ \text{cm}^{-3}}$$

$$E_f - E_i = 0.467\ \text{eV}$$

For the p-side since $N_a \gg n_i$ that means the doping dominates the overall number of carriers.
So, we can say, $p \cong N_a = 1 \times 10^{16}$ cm^{-3}
Using Eq. (3.22),

$$p_0 = n_i e^{(E_i - E_f)/kT}$$

Solving for $E_i - E_f$,

$$E_i - E_f = kT \times ln\frac{p_0}{n_i}$$

$$E_i - E_f = 0.0259\ \text{eV} \times ln\frac{1 \times 10^{16}\ \text{cm}^{-3}}{1.5 \times 10^{10}\ \text{cm}^{-3}}$$

$$E_i - E_f = 0.347\ \text{eV}$$

b)

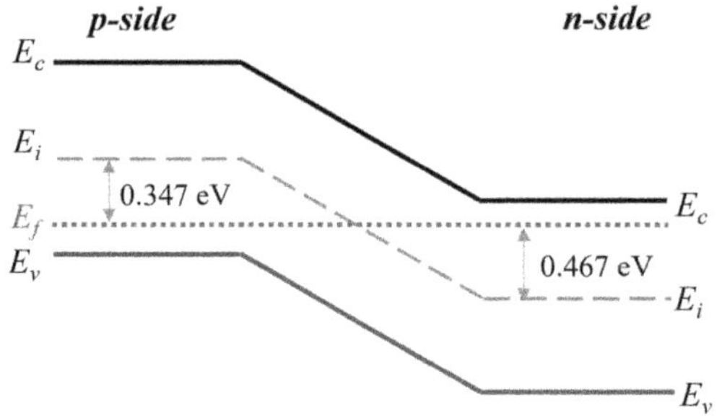

c) The drop in potential from the p-side to the n-side is the built-in potential V_0 from Eq. (4.1)

$$V_0 = 0.347 \text{ eV} + 0.467 \text{ eV} = 0.814 \text{ eV}$$

4.2.3 Calculation of the Built-In Potential

As shown in Example 4.1, the built-in potential can be determined from the band diagram. The built-in potential can also be calculated mathematically using balance of charge motion. Specifically, electron diffusion and drift currents must be equal since zero net current is flowing in the device. Similarly, the hole diffusion and drift currents match since the net current is zero. So, the total hole current can be written as

$$J_p = J_{p-drift} + J_{p-diffusion} = 0 \tag{4.3}$$

If we substitute from Eq. (3.69), then

$$q\mu_p p(x)\mathrm{E}_x - qD_p \frac{dp(x)}{dx} = 0 \tag{4.4}$$

and rearranging to

$$\frac{\mu_p}{D_p}\mathrm{E}_x = \frac{1}{p(x)}\frac{dp(x)}{dx} \tag{4.5}$$

Note that electrical field and potential are related by

$$\mathrm{E}(x) = \frac{dV(x)}{dx} \tag{4.6}$$

Further, by using Einstein relationship from Eq. (3.70), Eq. (4.5) becomes

$$\frac{1}{p(x)}\frac{dp(x)}{dx} = -\frac{q}{kT}\frac{dV(x)}{dx} \tag{4.7}$$

To solve this first-order linear differential equation we need to integrate both sides across the entire junction to calculate the potential drop, V. The potential at edge of the junction on the p-side is V_p and on the edge of the n-side is V_n. So, the limits for the integration of the voltage V will be V_p and V_n for the left side of Eq. (4.7). For the right-side integration limits for $p(x)$ are based on the hole and electron concentrations at the edge. Let us define p_p as the hole concentration on the p-side, and p_n is the hole concentration on the n-side. If we assume a step junction with regard to doping and one-dimensional function with x, then Eq. (4.7) with integration limits becomes

$$\int_{p_p}^{p_n} \frac{1}{p(x)} dp = -\int_{V_p}^{V_n} \frac{q}{kT} dV \tag{4.8}$$

The solution to this integral is

$$ln\frac{p_p}{p_n} = \frac{kT}{q}(V_n - V_p) \tag{4.9}$$

The hole concentration at the edge of the p-side is simply the doping concentration. Since $N_a \gg n_i$,

$$p_p = N_a \tag{4.10}$$

The hole concentration on the n-side can be calculated from Eq. (3.4). Since $n_0 p_0 = n_i^2$ and $n_n = N_d$, then

$$p_n = \frac{n_i^2}{N_d} \tag{4.11}$$

Substituting Eqs. (4.10) and (4.11) into Eq. (4.9) yields

$$ln\frac{N_a N_d}{n_i^2} = \frac{kT}{q}(V_n - V_p) \tag{4.12}$$

So, what does $V_n - V_p$ represent? From the physics part above, $V_n - V_p$ represents the built-in potential created from the fixed charges. So $V_n - V_p = V_0$.

$$V_0 = \frac{kT}{q} ln\frac{N_d N_a}{n_i^2} \tag{4.13}$$

Equation (4.13) allows us to determine the built-in voltage with the doping concentration on the p- and n-sides.

Example 4.2 ***Calculate and draw band diagram***
For the solar cell in Example 4.1

a) Calculate the built-in potential and compare it to 4.1 eV.
b) Repeat the calculation for a temperature of 400 K.

Solution

a) From Eq. (4.13)

$$V_0 = \frac{kT}{q} ln \frac{N_d N_a}{n_i^2}$$

$$V_0 = 0.0259 ln \frac{1 \times 10^{18} \times 1 \times 10^{16}}{(1.5 \times 10^{10})^2}$$

$$V_0 = 0.814 \text{ eV}$$

This matches the value of V_0 obtained from the band diagram in Example 4.1

b) For 400 K, same formula is used, Eq. (4.13), but with different parameters. First let us calculate $\frac{kT}{q}$ at 400 K

$$\frac{kT}{q}(400 \text{ K}) = 8.62 \times 10^{-5} (\text{eV/K}) \times 400 \text{ (K)}$$

$$\frac{kT}{q}(400 \text{ K}) = 0.0345 \text{ eV}$$

Next, we need to calculate n_i at 400 K using Eq. (3.5)

$$n_i(T) = BT^{3/2} e^{-E_g/2kT}$$

From Table 3.1 for Si, $B = 5.23 \times 10^{15}$ cm^{-3}K$^{-3/2}$

$$n_i(400) = 5.23 \times 10^{15} 400^{3/2} e^{-1.1/2(0.345)}$$

$$n_i(400) = 4.99 \times 10^{12} \text{ cm}^{-3}$$

You can also get value from Figure 3.5.

Now you can go back to use Eq. (4.13)

$$V_0(400 \text{ K}) = 0.0345 ln \frac{1 \times 10^{18} \times 1 \times 10^{16}}{(4.99 \times 10^{12})^2}$$

4.3 Junction Electrostatics

Let us take a closer look at the total charge in junction to calculate and plot the electric field and potential. This can also be used to determine the depletion width. Using the depletion approximation, only the fixed ionized impurities make up the charge in the junction. The mobile electrons and holes are quickly swept through. At equilibrium with the maximum depletion width, a peak electric field is reached to balance the diffusion. At that point, there are a fixed number of ionized impurities needed to reach the maximum electric field. The amount of negative charge must equal the amount of positive charge to contain the electrical field within the depletion width W and to maintain charge neutrality. Also, since the electric field lines must start at a positive charge and end at a negative charge, Figure 4.9 shows the *pn* junction highlighting the charge, where the W is the maximum depletion width. We can also define x_n as the depletion on the n-side and x_p as the depletion on the p-side.

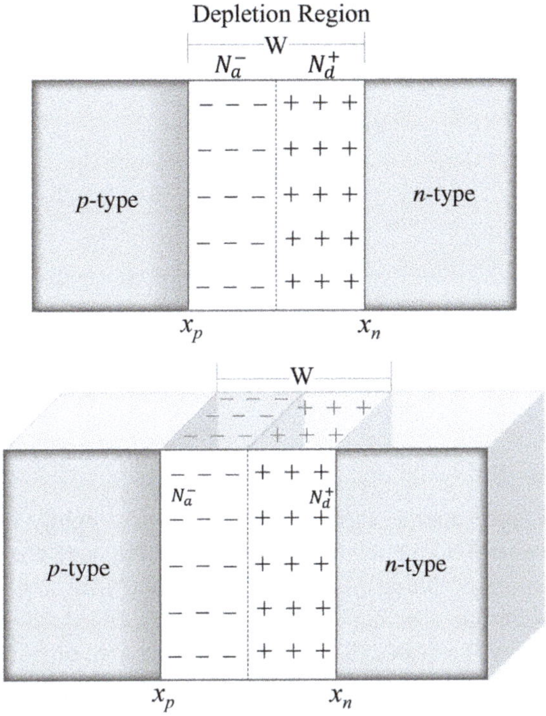

Figure 4.9 A *pn* junction highlighting the charge within the depletion region *W* and the depletion on the *n*-side (x_n) and the depletion on the *p-side* (x_p) illustrated in (a) 2D and (b) 3D.

It is important to remember these structures are three dimensional so they need to be calculated in terms of volume. It is helpful analogy to think of it like slicing a piece of cake. The total charge in that slice of cake is the volume of the slice multiplied by the charge per volume. In a semiconductor, the volume of the slice is multiplied by the depletion depth in the *n* (x_n) and the *p* (x_p) regions.

Let us define Q_n as total charge on the *n*-side and Q_p as total charge on the *p*-side. As shown in Figure 4.9, Q_n will be positive due to the N_d^+ and Q_p is negative due to the N_a^- ionized impurities. Assuming 100% doping ionization, the total charge on either side of the junction is given by

$$Q_n^+ = qAx_n N_d \tag{4.14}$$

$$Q_p^- = qAx_p N_a \tag{4.15}$$

For charge neutrality we have that $|Q_p| = |Q_n|$.

That means with complete ionization, the charge neutrality gives

$$x_n N_d = x_p N_a \tag{4.16}$$

The charge at the junction can be plotted across the junction in Figure 4.10.

One important conclusion from this formula is that the depletion region extends further into the lightly doped side of the junction. This is intuitive

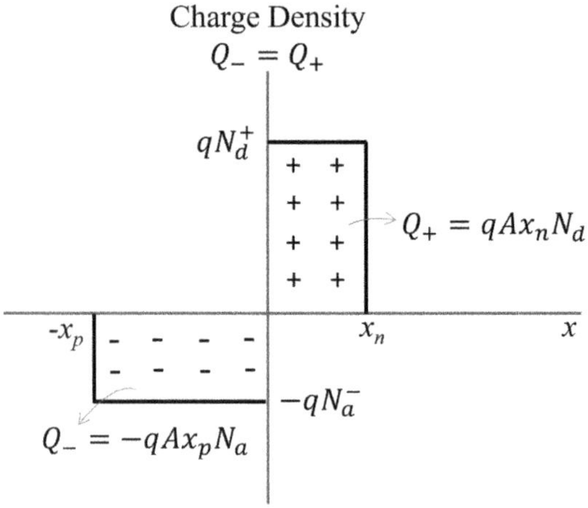

Figure 4.10 Illustration of the charge density in the depletion region of a *pn* junction.

because charge carriers diffusing into a lightly doped region must penetrate deeper before encountering sufficient opposite charge to recombine, in order to maintain charge neutrality. Equivalently, from a volume perspective, a larger region must be depleted on the lightly doped side so that enough fixed ionized charge is exposed to balance the total charge on both sides of the junction.

Another important thing to determine is the electric field in the junction. This can be done by using Poisson's equation for electrostatics that relates the slope of the electric field to the charge at any point in the junction. To simplify the calculation, the depletion approximation is used and n and p regions are calculated separately. Let us take a Gaussian sphere in the depletion region to calculate the total charge in the n and p regions. Figure 4.11 shows the Gaussian sphere used. Since the mobile charges are assumed to be zero, then the total charge is only due to the fixed charge from the doping.

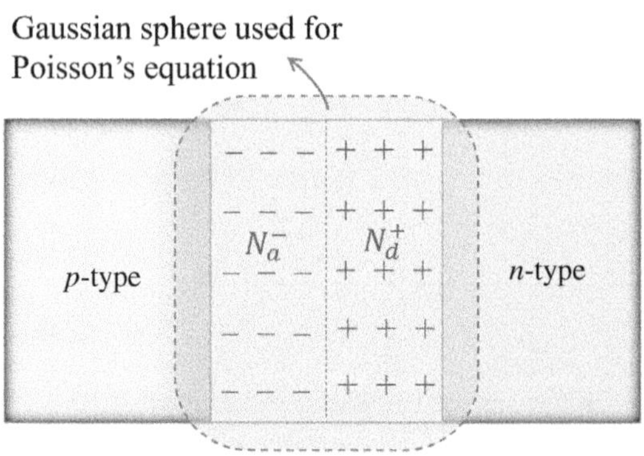

Figure 4.11 Illustration of Gaussian sphere used to calculate total charge inside depletion region.

Table 4.1 Dielectric constants for Si, Ge, and GaAs.

	Dielectric constant ε_r
Si	11.7
Ge	13.1
GaAs	16.0

Then, Poisson's equation gives

$$\frac{q}{\varepsilon}\left(N_d^+ - N_a^-\right) = \frac{d\mathrm{E}(x)}{dx} \tag{4.17}$$

Here, ε is $\varepsilon_0\varepsilon_r$, where ε_0 is the permittivity in a vacuum with value 8.85×10^{-14} F/cm and ε_r is the relative permittivity of the material. Table 4.1 shows the relative permittivity of some key semiconductor materials.

Assuming complete ionization of the total charge, while treating both sides separately and assuming uniform step like doping, we then have

$$\frac{q}{\varepsilon}(N_d) = \frac{d\mathrm{E}(x)}{dx} \quad (n\text{-side}) \tag{4.18}$$

$$\frac{q}{\varepsilon}(N_a) = -\frac{d\mathrm{E}(x)}{dx} \quad (p\text{-side}) \tag{4.19}$$

On the n-side, the solution to this formula is the integral

$$\mathrm{E}(x) = \int_0^x \frac{q}{\varepsilon}N_d\, dx = \frac{q}{\varepsilon}N_d x + C \tag{4.20}$$

We can solve for the constant of integration by analyzing the physics of junction. Based on the depletion approximation, electric field is zero ($E = 0$) outstide the depletion region. This can be the boundary condition to solve the integration problem. Specifically, at $x = x_n$, the electric field will drop to 0. So, using Eq. (4.18), we get

$$\mathrm{E}(x_n) = \frac{q}{\varepsilon}N_d x_n + C = 0 \tag{4.21}$$

So, C is

$$C = -\frac{q}{\varepsilon}N_d x_n \tag{4.22}$$

And we can write Eq. (4.20) as

$$\mathrm{E}(x) = \frac{q}{\varepsilon}N_d(x - x_n) \tag{4.23}$$

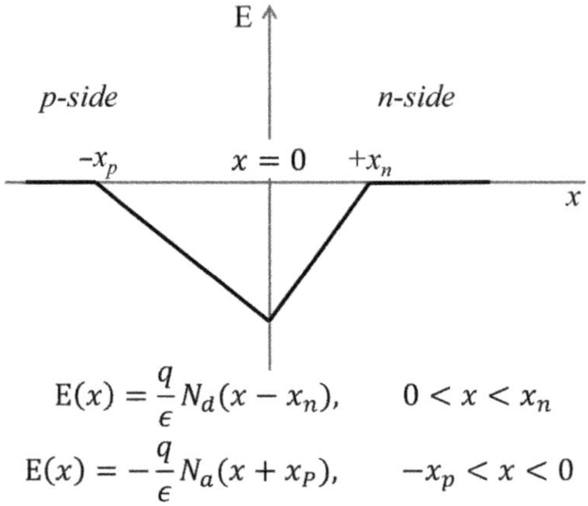

Figure 4.12 Electric field, *E*, of the junction as a function of distance *x*.

This same analysis can be done for the *p*-side starting from Eq. (4.16).

$$\mathrm{E}(x) = \int_{-x}^{0} -\frac{q}{\varepsilon} N_a \, dx = -\frac{q}{\varepsilon} N_a x + B \tag{4.24}$$

As before, when $x = -x_p$, then $\mathrm{E}(x_p) = 0$.

$$\mathrm{E}(x_p) = -\frac{q}{\varepsilon} N_d x_p + B = 0 \tag{4.25}$$

So, B is

$$B = \frac{q}{\varepsilon} N_a x_p \tag{4.26}$$

And we can write Eq. (4.26) as

$$\mathrm{E}(x) = -\frac{q}{\varepsilon} N_a (x + x_P) \tag{4.27}$$

The maximum electric field can be found, when $x = 0$, either from Eq. (4.23) or Eq. (4.27).

$$\mathrm{E}(0) = \mathrm{E}_0 = \mathrm{E}_{\max} = -\frac{q}{\varepsilon} N_d x_0 = -\frac{q}{\varepsilon} N_a x_{p0} \tag{4.28}$$

Both equations also show that the electric field is negative across the entire junction. This is consistent with the physics since the electric field lines point from *n* to *p* (from N_d^+ to N_a^-).

Figure 4.12 plots the electric field of the junction vs. x highlighting all the key points.

From the electric field, the potential can be determined using the following relationship:

$$\mathrm{E} = \frac{dV}{dx} \tag{4.29}$$

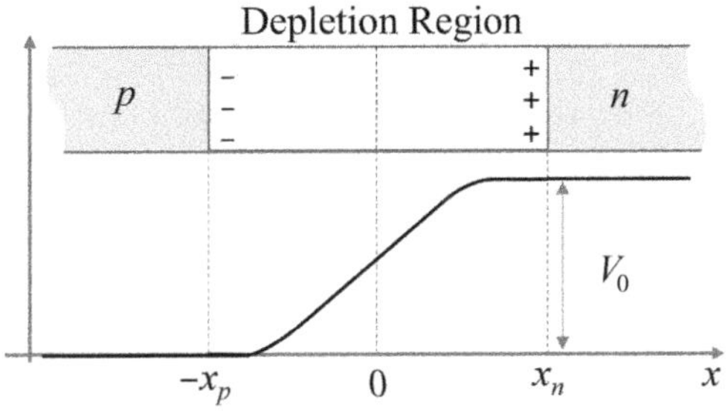

Figure 4.13 Energy potential across the *pn* junction. Also shown are *p*-, *n*-, and depletion regions for reference.

V is the integral of the electrical field and since $E(x)$ is linear, we have that

$$V(x) \propto x^2 \tag{4.30}$$

Figure 4.13 shows the potential across the junction.

The total negative contact potential can be calculated using the integral across the junction,

$$-V_0 = \int_{x_p}^{x_n} \mathrm{E}(x)dx \tag{4.31}$$

Next, we can calculate the peak potential as the maximum area under the curve. Since $E(x)$ has a triangular shape, using formula ½ base × height for the area of triangle gives

$$-V_0 = \frac{1}{2}W\mathrm{E}_0 \tag{4.32}$$

Here, W is the total depletion width $x_n + x_p$.

We can now substitute E_0 from Eq. (4.28) into Eq. (4.32). Either the *n*- or *p*-side:

$$V_0 = -\frac{1}{2}W\frac{q}{\varepsilon}N_d x_{n0} = \frac{1}{2}W\frac{q}{\varepsilon}N_a x_{p0} \tag{4.33}$$

Using Eq. (4.16) and the fact that $W = x_n + x_p$, then

$$x_n = W\left(\frac{N_a}{N_a + N_d}\right) \tag{4.34}$$

and

$$x_p = W\left(\frac{N_d}{N_a + N_d}\right) \tag{4.35}$$

Substituting Eq. (4.28) into Eq. (4.33) yields

$$V_0 = \frac{1}{2}\frac{q}{\varepsilon}\frac{N_a N_d}{N_a + N_d}W^2 \tag{4.36}$$

Solving for W in terms of V_0 gives

$$W = \sqrt{\frac{2\varepsilon(V_0 - V_a)}{q}\left(\frac{1}{N_a} + \frac{1}{N_d}\right)} \tag{4.37}$$

In Eq. (4.37), we introduce the applied voltage (V_a) and set it equal to 0 for now. The effect of V_a will be discussed in the next section. Setting $V_a = 0$ and replacing V_0 in Eq. (4.37) with its formula as a function of doping from Eq. (4.13), we obtain Eq. (4.38), which is W as a function of doping,

$$W = \sqrt{\frac{2\varepsilon kT}{q^2} ln \frac{N_a N_d}{n_i^2} \left(\frac{1}{N_a} + \frac{1}{N_d}\right)} \tag{4.38}$$

This formula allows us to calculate the depletion region width of a *pn* junction with the doping concentration on both sides at zero applied bias.

Example 4.3 ***Calculate and sketch* pn *junction***

Take a Si *pn* junction in Example 4.1. The junction has a square area with side 0.25 cm.

a) Calculate W, x_n, x_p
b) Calculate E_0 and plot $E(x)$
c) Calculate Q_n^+

Solution

a) Using Eq. (4.37) with $V_a = 0$

$$W = \sqrt{\frac{2\,\varepsilon_o \varepsilon_r\, V_0}{q} \left(\frac{1}{N_a} + \frac{1}{N_d}\right)}$$

$$W = \sqrt{\frac{(2)(8.5 \times 10^{-14})(11.8)\,(0.814)}{1.6 \times 10^{-19}} \cdot \left(\frac{1}{1 \times 10^{18}} + \frac{1}{1 \times 10^{16}}\right)}$$

$$W = 3.38 \times 10^{-5}\ \text{cm} = 0.338\ \mu\text{m}$$

Using Eq. (4.35),

$$x_n = W\ \frac{N_d}{N_a + N_d} = 3.38 \times 10^{-5}\ \frac{1 \times 10^{18}}{1 \times 10^{16} + 1 \times 10^{18}}\ \text{cm}$$

$$x_n = 3.35 \times 10^{-5}\ \text{cm} = 0.335\ \mu\text{m}$$

Notice that the depletion width extends mostly to the lightly doped *n* side. To find the x_p, use $W = x_n + x_p$ or Eq. (4.35)

$$x_p = W\ \frac{N_a}{N_a + N_d} = 3.38 \times 10^{-5}\ \text{cm}\ \frac{1 \times 10^{16}}{1 \times 10^{16} + 1 \times 10^{18}}$$

$$x_p = 3.34 \times 10^{-7}\ \text{cm} = 3.4 \times 10^{-4}\ \mu\text{m}$$

To check $W = x_n + x_p = 3.35 \times 10^{-5}\ \text{cm} + 3.34 \times 10^{-7}\ \text{cm} = 3.38 \times 10^{-5}\ \text{cm}$

b) Calculate E_{max} and plot $E(x)$

Using Eq. (4.28), E_{max} can be calculated using the n or p-side

$$E_{max} = -\frac{q}{\varepsilon} N_d x_0$$

$$E_{max} = -\frac{1.6 \times 10^{-19}\ C}{(8.5 \times 10^{-14})(11.8)} (1 \times 10^{16}\ cm^{-3})(3.35 \times 10^{-5}\ cm)$$

$$E_{max} = -5.13 \times 10^4\ \ V/cm$$

The negative sign represents the built-in electric field pointing from the n to p-side

Plot of $E(x)$:

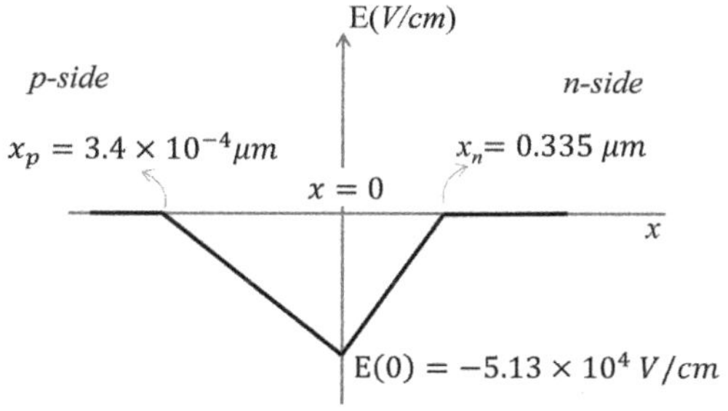

c) To find the Q_n^+, use Eq. (4.14)

$$Q_n = qAx_n\, N_d$$

$$Q_n^+ = (1 \times 10^{-19})(0.25^2\ cm^2)(3.35 \times 10^{-5}\ cm)(1 \times 10^{16}\ cm^{-3})$$

$$Q_n^+ = 2.093 \times 10^{-9}\ C$$

4.4 *pn* Junction with Applied Voltage

From circuit theory, a diode is shown to have an asymmetric IV curve with an exponentially increasing current on one side and very little constant current on the other side. To the contrary, the resistor follows ohms law with formula $V = IR$ and hence is linear. Figure 4.14 shows this ideal diode IV curve compared to a resistor.

In this section, the effect of applying a voltage and the physics of why the current in the diode is asymmetric with voltage is explained. The effect of voltage is discussed in terms of forward and reverse bias conditions. In reverse bias, the potential on the p-side increases relative to the n-side. In forward bias the potential on the p-side is reduced relative to the n-side. Using this physics, the reason for asymmetric behavior is explained and mathematically quantified.

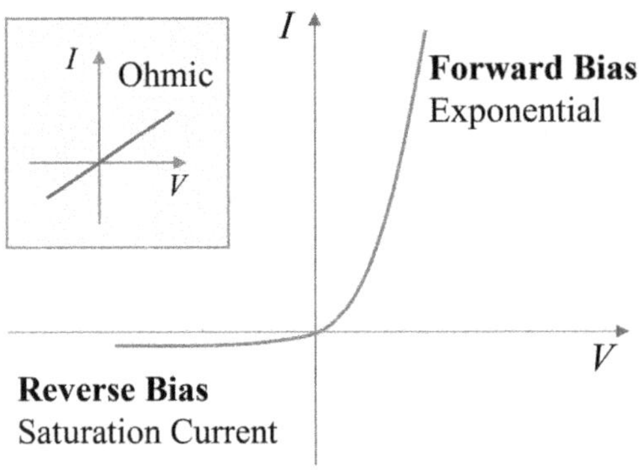

Figure 4.14 Diode IV characteristics. Inset shows ohmic IV curve of a resistor.

4.4.1 Reverse Bias

In reverse bias the *pn* junction is connected to an external battery, with the negative terminal on the *p*-side and the positive on the *n*-side. Figure 4.15 shows a photo of the reverse bias connection with a negative applied voltage ($V_a < 0$).

One assumption to simplify the physics is to assume the voltage applied will drop only in the depletion region and not in the regions outside. This assumption is valid since the resistance is small outside the depletion region. Applying a negative voltage to the *p*-side relative to the *n*-side has the effect of increasing the potential barrier between the *n*- and *p*-sides. With no voltage bias applied, the barrier between the *p*- and *n*-sides is built-in potential, V_o, as shown earlier. When a reverse voltage is applied, the built-in voltage increases by the amount of the applied reverse bias. The new barrier with a reverse bias (V_{RB}) is given by

$$V_{RB} = V_0 + |V_a| \tag{4.39}$$

Figure 4.16 shows the band diagram of the *pn* junction in thermal equilibrium and with reverse bias, showing the barrier increase. Since the electric field increases in the junction with reverse bias, the width of the depletion region must

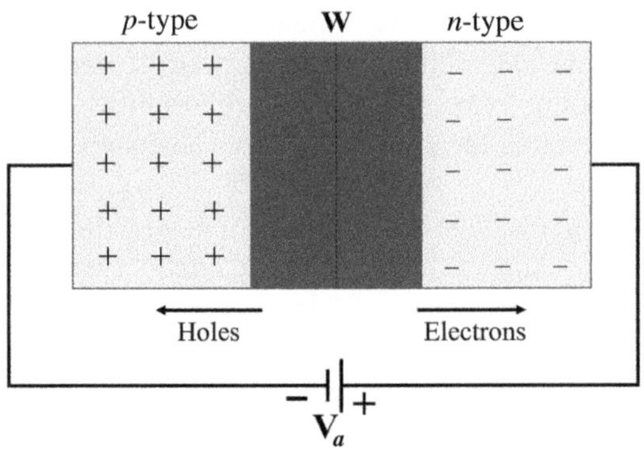

Figure 4.15 Illustration of reverse bias connection.

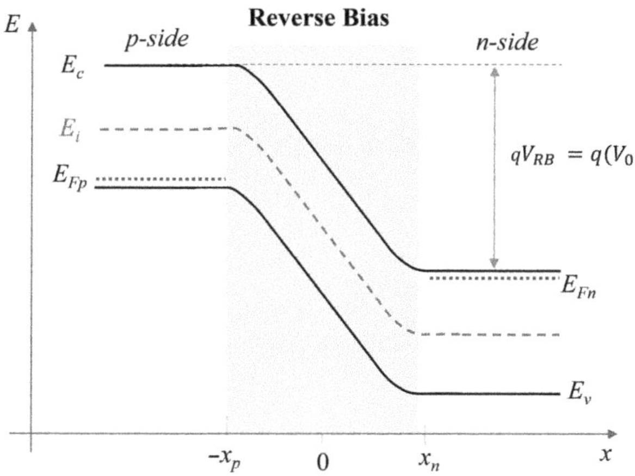

Figure 4.16 The band diagram of the solar cell in thermal equilibrium and with reverse bias showing the barrier increase.

increase due to the increase in ionized dopant atoms as shown in Figure 4.15. This is consistent with Eq. (4.37); when $V_a < 0$, W will increase with applied reverse bias.

Based on Figure 4.16 the large barrier will affect the motion of carriers. This essentially cuts off all diffusion of carriers. However, from the IV curve shown in Figure 4.14, a small negative current is measured in reverse bias. What is the physics of this negative current? Let us start with the *p*-side. Thermally generated carriers in a semiconductor create EHPs far from the junction and do not impact the current since they recombine before they reach the built-in electric field. However, what about electrons that are created close to the junction? Can they make it there before recombination? Yes, mathematically speaking, if the electron is created within a diffusion length, there is a very high probability that they can make it to the junction and drift to the other side by the electric field. These electrons move from *p*- to *n*-side and provide a current directed from *n* to *p* that is negative. This same physics can be explained on the *n*-side. The holes that are generated near the junction, within a hole diffusion length, can be swept from *n*- to *p*-side. These holes add a negative current from *n*- to *p*-side, adding to the electron current component. So, the total current under reverse bias will be the holes and electrons that are thermally generated near the junction and are swept by the electric field to the other side. Figure 4.17 shows the motion of carriers in reverse bias on a band diagram.

This current is sometimes called a generation current since the main origin is the generation of carriers near the junction. After the generation, the carriers diffuse and then drift to the electric field. From Figure 4.14, the IV curve of the diode

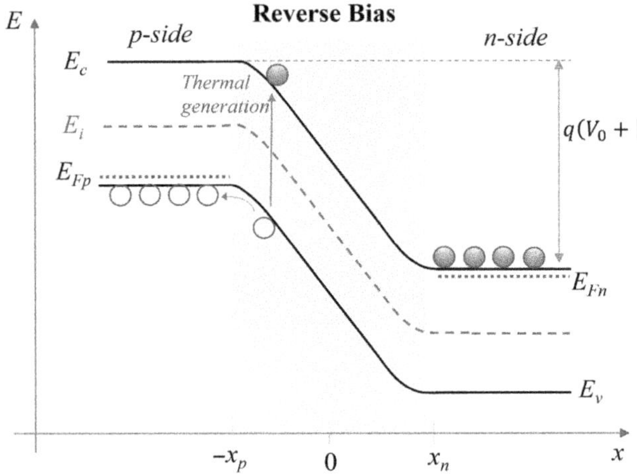

Figure 4.17 Motion of carriers under reverse bias on a band diagram.

shows that the current in reverse bias does not change with voltage. At first glance, it looks like the higher the voltage, the higher electric field, and the more drift current. So why is the current not changing with voltage? From the band diagram, it shows that the drift is not the origin of the reverse bias current. The origin is the generation of EHPs within a depletion length. The amount of generation is not affected by the electric field and only a function of doping. The value of the electric field will only change how fast the carriers are swept across the junction, but not how many make it. This is the reason the reverse bias current does not change with bias and is constant, which is rather counterintuitive.

Example 4.4 ***Effect of applied reverse bias***
Repeat Example 4.3 but with $V_a = -3$ V.

a) Using Eq. (4.37) with $V_a = -3$ V

$$W = \sqrt{\frac{2\,\varepsilon_o \varepsilon_r\,(V_0 - V_a)}{q}\left(\frac{1}{N_a} + \frac{1}{N_d}\right)}$$

$$W = \sqrt{\frac{(2)(8.5\times10^{-14})(11.8)\,(0.814+3)}{1.6\times10^{-19}}\cdot\left(\frac{1}{1\times10^{18}} + \frac{1}{1\times10^{16}}\right)}$$

$$W = 6.9\times10^{-5}\ \text{cm} = 0.69\ \mu\text{m}$$

Using Eq. (4.35),

$$x_n = W\,\frac{N_d}{N_a + N_d} = 6.9\times10^{-5}\,\frac{1\times10^{18}}{1\times10^{16} + 1\times10^{18}}\ \text{cm}$$

$$x_n = 6.88\times10^{-5}\ \text{cm} = 0.688\ \mu\text{m}$$

Notice that the depletion width extends mostly to the lightly doped n side.
To find the x_p, use $W = x_n + x_p$

$$x_p = W \frac{N_a}{N_a + N_d} = 6.9 \times 10^{-5}\ \text{cm} \frac{1 \times 10^{16}}{1 \times 10^{16} + 1 \times 10^{18}}$$

$$x_p = 6.9 \times 10^{-7}\ \text{cm} = 6.9 \times 10^{-3}\ \mu\text{m}$$

To check $W = x_n + x_p = 6.88 \times 10^{-5}\ \text{cm} + 6.9 \times 10^{-7}\ \text{cm} = 6.9 \times 10^{-5}\ \text{cm}$

b) Calculate E_{max} and plot $E(x)$
Using Eq. (4.28), E_{max} can be calculated using the n or p-side

$$E_{max} = -\frac{q}{\varepsilon} N_d x_n$$

$$E_{max} = -\frac{1.6 \times 10^{-19}\ \text{C}}{(8.5 \times 10^{-14})(11.8)} (1 \times 10^{16}\ \text{cm}^{-3})(6.88 \times 10^{-5}\ \text{cm})$$

$$E_{max} = -1.1 \times 10^{5}\ \text{V/cm}$$

The negative sign represents the built-in electric field pointing from the n to p-side
Plot of $E(x)$:

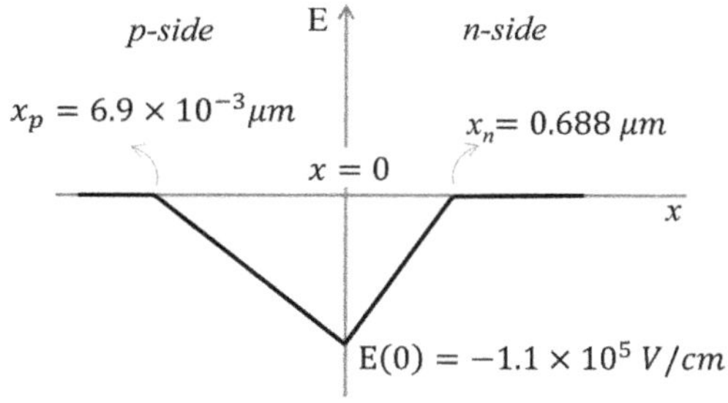

c) To find the Q_n^+, use Eq. (4.14)

$$Q_n = qAx_n N_d$$

$$Q_n^+ = (1 \times 10^{-19})(0.25^2\ \text{cm}^2)(6.88 \times 10^{-5}\ \text{cm})(1 \times 10^{16}\ \text{cm}^{-3})$$

$$Q_n^+ = 6.9 \times 10^{-9}\ \text{C}$$

4.4.2 Forward Bias

In the other case, a positive voltage is applied to the p-side and negative voltage to the n-side. This case is called forward bias. Figure 4.18 shows the forward bias connection with $V_a > 0$.

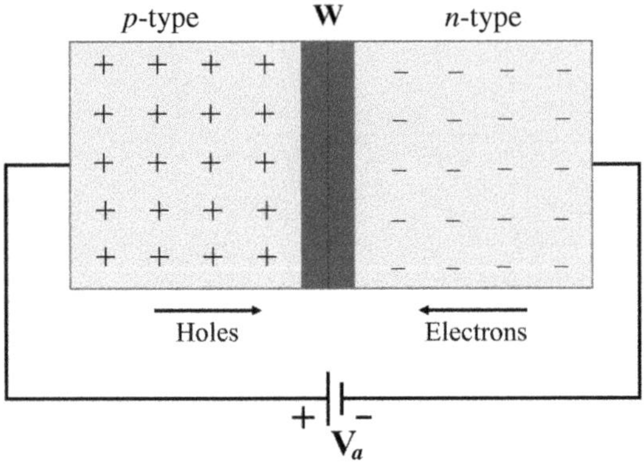

Figure 4.18 Illustration of forward bias connection.

The potential of the p-side relative to the n-side is reduced. This reduces the barrier for electrons on the n-side facing the p-side. It also reduces the barrier for holes from p-side to n-side. Remember, in equilibrium, the barrier maintains the equilibrium by reducing the diffusion of carriers from n to p and p to n. Once the barrier is reduced, the floodgates open, allowing for diffusion of majority carriers to either side. The amount of carriers that diffuse will increase exponentially due to the potential barrier. This means only small changes in barrier height can provide large changes in the carrier concentration. The exponential dependence on the number of carriers that diffuse explains why the current in forward bias is exponentially increasing with voltage. Electrons diffuse from n to p and holes from p to n with forward bias. When a forward voltage is applied the built-in voltage decreases by the amount of the applied forward bias. The new barrier with a forward bias (V_{FB}) is given by

$$V_{FB} = V_0 - |V_a| \tag{4.40}$$

Figure 4.19 shows the band diagram under the forward bias.

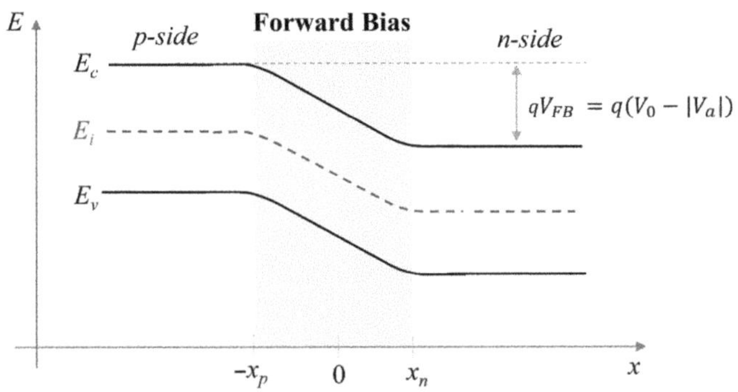

Figure 4.19 Band diagram under the forward bias.

4.4.3 Ideal Diode Equation

Based on the previous band diagrams, the physics of the diode current mechanisms are understood. The forward bias barrier is reduced, allowing for exponential amount of diffusion current. In reverse bias, the barrier is large, and only generation current is measured to quantify this mathematically by modeling it as carrier injection problem. Specifically, electrons are injected from the *n*- to *p*-side. Once on the *p*-side they are the minority carrier, and they diffuse and recombine, providing an electron diffusion current there. The holes injected from *p*- to *n*-side become the minority carriers on the *n*-side. They also diffuse and recombine, making a hole current. The formula that describes the diode current is

$$I = I_0 \left(e^{qV_a/kT} - 1\right) \tag{4.41}$$

where V_a is the applied voltage either positive for forward bias or negative for reverse bias and I_0 is the constant reverse bias saturation current. In forward bias, when V_a is positive, the exponential term will dominate and that explains the large current. When V_a is negative, the exponential term is very small, and the current is $-I_0$, the reverse saturation current. A formula for the reverse saturation current with key physical parameters is

$$I_0 = qA\left(\frac{D_p}{L_p}p_n + \frac{D_n}{L_n}n_p\right) \tag{4.42}$$

The equation is divided into *p*- and *n*-sides. The *n*-side deals with the minority carrier holes. The *p*-side with the minority carrier electrons. The subscript in *p* and *n* helps determine if it is *n*- or *p*-side. The formula includes the amount of the minority carrier electrons and holes generated, the diffusion length, and diffusivity. Figure 4.20 shows the semi-log *IV* plot of Eq. (4.40).

Example 4.5 ***Calculate current in* pn *junction***
Take a *Ge pn* junction at 300 K doped $N_d = 5 \times 10^{18}$ cm^{-3}, $N_a = 1 \times 10^{16}$ cm^{-3}. The junction has a square area with side 0.5 cm^2.
n-side:

$\mu_n = 2500$ cm^2/V · s and $\mu_p = 400$ cm^2/V · s
$\tau_n = 5$ ns and $\tau_p = 1$ ns

p-side:

$\mu_n = 3900$ cm^2/V · s and $\mu_p = 1500$ cm^2/V · s
$\tau_n = 20$ ns and $\tau_p = 4$ ns

a) Calculate the reverse saturation current I_0
b) Calculate the current at a voltage of 0.5 V

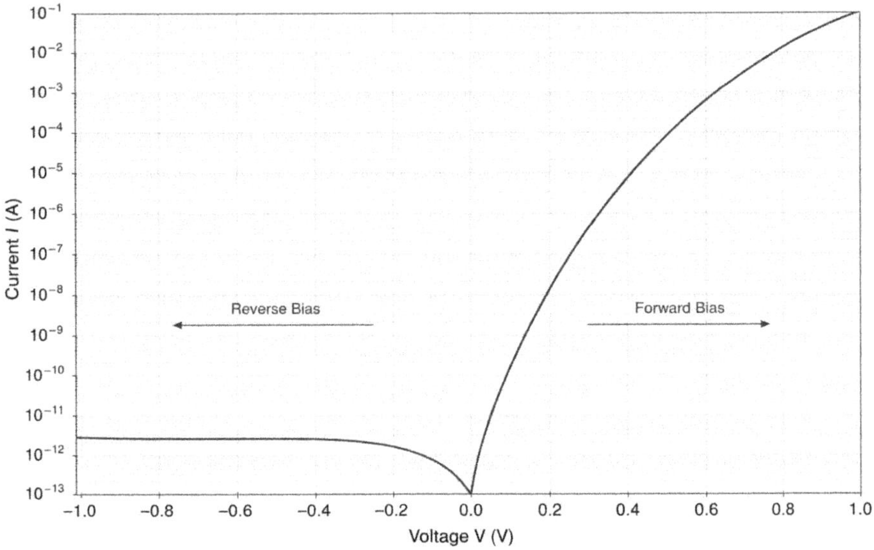

Figure 4.20 Example of current–voltage semi-log plot. (Image generated using OpenAI, ChatGPT).

Solution

Use Eq. (4.42):

$$I_0 = qA\left(\frac{D_p}{L_p}p_n + \frac{D_n}{L_n}n_p\right)$$

n-side:

Using Eq. (3.66),

$$\frac{D_p}{\mu_p} = \frac{kT}{q} \text{ so } D_p = 400\ (\text{cm}^2/\text{V}\cdot\text{s})\ 0.0259\ (\text{eV}) = 10.36\ \text{cm}^2/\text{s}$$

Using Eq. (3.68),

$$L_p = \sqrt{D_p\tau_p},$$

$$L_p = \sqrt{400\ (\text{cm}^2/\text{V}\cdot\text{s}) \times 1 \times 10^{-9}\ (\text{s})} = 6.32 \times 10^{-4}\ \text{cm} = 632\ \mu\text{m}$$

p_n is the minority carrier hole concentration on the *n*-side. Since $N_d \gg n_i$ then $n_0 = N_d = 5 \times 10^{18}\ \text{cm}^{-3}$, so

$$p_0 = p_n = \frac{n_i^2}{N_d} = \frac{(1.5 \times 10^{10})^2}{5 \times 10^{18}} = 45\ \text{cm}^{-3}$$

p-side:

Using Eq. (3.66),

$$\frac{D_n}{\mu_n} = \frac{kT}{q} \text{ so } D_n = 1500\,(\text{cm}^2/\text{V}\cdot\text{s}) \times 0.0259\,(\text{eV}) = 38.85\,\text{cm}^2/\text{s}$$

Using Eq. (3.70),

$$L_n = \sqrt{D_n \tau_n},\quad L_n = \sqrt{1500\,(\text{cm}^2/\text{V}\cdot\text{s}) \times 20 \times 10^{-9}(\text{s})}$$
$$= 5.48 \times 10^{-4}\,\text{cm} = 548\,\mu\text{m}$$

n_p is the minority carrier electron concentration on the *p*-side. Since $N_a \gg n_i$, $p_0 = N_a = 1 \times 10^{16}\,\text{cm}^{-3}$, and then

$$n_0 = n_p = \frac{n_i^2}{N_a} = \frac{(1.5 \times 10^{10})^2}{1 \times 10^{16}} = 22\,500\,\text{cm}^{-3} = 2.25 \times 10^4\,\text{cm}^{-3}$$

Area of the device is

$$A = 0.5\,\text{cm} \times 0.5\,\text{cm} = 0.25\,\text{cm}^2$$

Now we plug into Eq. (4.41)

$$I_0 = qA\left(\frac{D_p}{L_p} p_n + \frac{D_n}{L_n} n_p\right)$$

$$I_0 = 1.6 \times 10^{-19}\,\text{C} \times 0.25\,\text{cm}^2 \left(\frac{10.36}{6.32 \times 10^{-4}} 45 + \frac{38.85}{5.48 \times 10^{-4}} 2.25 \times 10^4\right)$$

$$I_0 = 6.32 \times 10^{-11}\,\text{A}$$

b) To find the current at $V_a = 0.5$ V, we can use Eq. (4.41)

$$I = I_0\left(e^{qV_a/kT} - 1\right)$$

$$I = 6.32 \times 10^{-11}(e^{0.5/0.0259} - 1)\,\text{A}$$

With $kT = 0.0259$ eV at 300 K

$$I = 6.32 \times 10^{-11}(e^{0.5/0.0259} - 1)$$

$$I = 15.5\,\text{mA}$$

4.5 Nonideal Diode

So far, the diode was treated as an ideal diode. The ideal diode does not account for some second-order effects. These include reverse bias breakdown voltage, the effect of built-in potential on turn-on voltage, and recombination/generation in the depletion region.

4.5.1 *pn* Junction Breakdown

The ideal *pn* junction theory predicts that the negative reverse bias current does not change with voltage and is very small. It turns out that with large negative applied reverse bias, the current will suddenly increase. Figure 4.21 shows an *IV* curve with breakdown.

This is called the reverse bias breakdown. It should be pointed out this is not an actual breakdown as the device still operates and can be used in this regime. So, what is the physics that causes this increase in negative current? There are two physical mechanisms that cause this sudden increase: impact ionization and quantum mechanical tunneling.

Let us assume the *pn* junction is lightly doped. As the reverse bias voltage continues to increase, the electric field increases and the depletion width increases. At such high electric fields, electrons can reach very high velocity, gaining high kinetic energy. Lattice scattering events or collisions from this high-velocity electrons release energy, and this energy is absorbed in the Si lattice. The collision energy can be large enough to break a $Si - Si$ covalent bond allowing for EHP generation (electron in the conduction band and hole in the balance band). This process is called impact ionization. This impact ionization process triggers an avalanche effect, with the new electrons also moving with high velocities, colliding, and making new EHPs. The new carriers are moving by a drift in the electric field. This explains the sudden increase in negative current. An analogy is an avalanche snow effect where the snowball continues to get bigger as it rolls down a mountain. In this case, more new electrons and holes are created by impact ionization with multiplication effect. Figure 4.22 illustrates impact ionization effect on the band diagram.

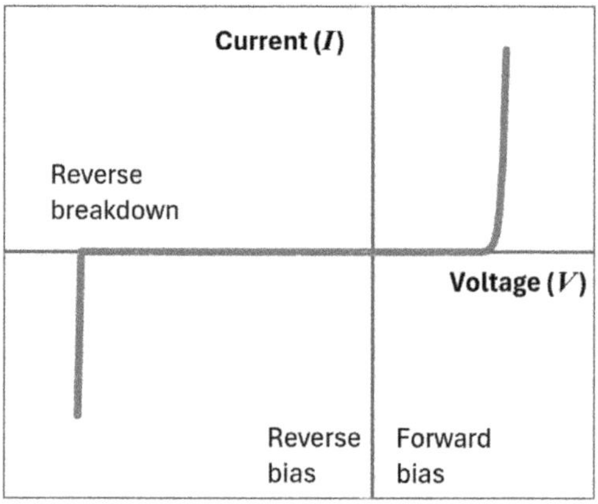

Figure 4.21 Current–voltage characteristics with reverse voltage breakdown.

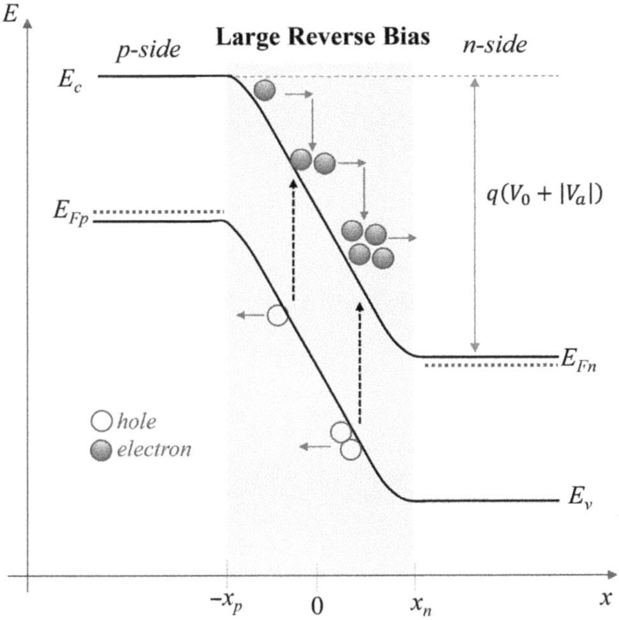

Figure 4.22 Illustration of impact ionization process.

In addition to impact ionization, quantum mechanical tunneling can lead to reverse saturation breakdown. Let us take the case, the *p*- and the *n*-side are heavily doped and draw the band diagram.

On the reverse bias, the valence band of the *p*-side can line up directly with the conduction band of the *n* side. This is the perfect condition for tunneling. Quantum mechanical tunneling utilizes the nature of electrons being both charged particles and waves. Tunneling is where charged particles can pass through a barrier. Unlike in classical physics, where a particle needs enough energy to go over a barrier, quantum mechanics allows particles to "tunnel" through it due to their wave-like nature. This effect is crucial in many areas of physics and technology, such as nuclear fusion. These electrons and holes that make it are moving in from *p* to *n*, which explains the large negative current seen in the diode *IV* curve. Figure 4.23 shows the band diagram of the tunneling effect.

This diode is called a Zener diode after the American theoretical physicist Clarence Zener who first described the quantum mechanical electrical breakdown phenomenon in 1934. Zener diodes are used for voltage regulation, as reference elements, surge suppressors, and in switching applications and clipper circuits.

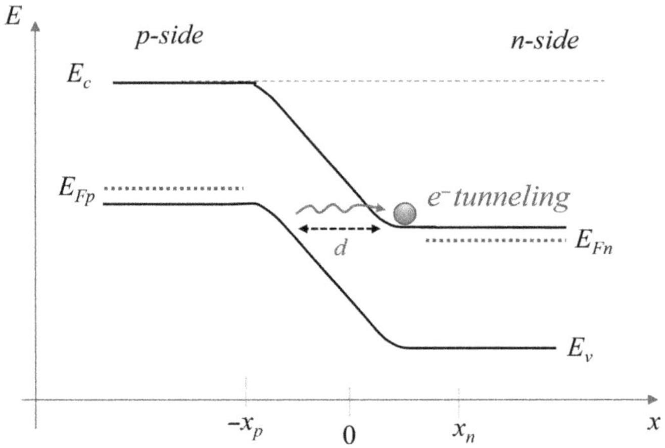

Figure 4.23 Illustration of band diagram with the tunneling effect.

4.5.2 Turn-on Voltage of Diode

In the ideal diode, the current exponentially begins to increase exactly at $V=0$. However, experimentally measured data shows that there is a turn-on voltage greater than 0. Figure 4.24 shows measured diodes with different semiconductor materials.

We notice from data that the bandgap is strongly affecting the turn-on voltage. If Eq. (4.42) is substituted into Eq. (4.10), then

$$I = \left(\frac{D_p}{L_p}p_n + \frac{D_n}{L_n}n_p\right)\left(e^{-qV_a/kT} - 1\right) \tag{4.43}$$

For the case when the hole concentration is much larger than the electron concentration, called a p^+n diode, and for $V_a \gg kt$, we have that

$$I = \left(\frac{D_p}{L_p}p_n\right)\left(e^{-qV_a/kT}\right) \tag{4.44}$$

If we substitute Eq. (3.15) for p_n, then

$$I = \left(\frac{D_p}{L_p}N_v\right)\left(e^{[qV_a-(E_{fn}-E_{fv})/kT]}\right) \tag{4.45}$$

From this formula, we can see if the forward bias is much smaller than the separation of the quasi-Fermi levels, the current remains very small. In other words, before the exponential current starts, a minimum voltage is needed based on this separation. Specifically, the separation upper limit is the contact potential as shown in the band diagram. This is the reason current increases rapidly, at

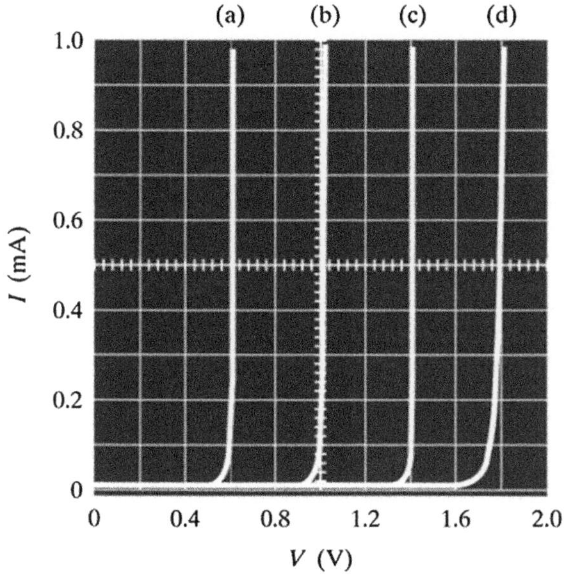

Figure 4.24 Example of practical *IV* curves for diodes made of materials with various bandgap values. *Source:* [1] / with permission of Pearson Education.

voltages less than bandgap. This effect was not part of the ideal diode since the change in the majority carriers was not included.

4.5.3 Recombination Generation in the Depletion Region

In the ideal diode analysis, the generation and recombination in the depletion region were not included. This generation and recombination can be enhanced with trap states in the junction. The band diagram in Figure 4.25 shows the depletion region with the trap states at mid-gap.

In forward bias, as electrons and holes overcome the barrier, they will move in the depletion region. Although the time spent in the depletion region is short, carriers can recombine as they overcome the barrier to travel into the depletion region. This effect is only reduced if the width of the depletion region is small compared to the diffusion length. Mathematically, it is difficult to calculate the recombination current since it depends on trap and carrier concentration with location. However, the effect of recombination in the depletion region can be modeled simple by including non-ideality factor n in the exponential part of the formula

$$I = I_0(e^{qV/\mathrm{nk}T} - 1) \tag{4.46}$$

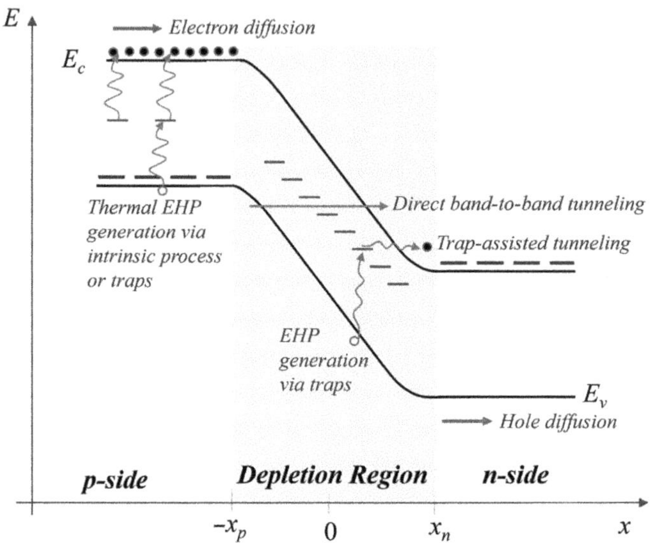

Figure 4.25 Band diagram showing depletion region with trap states and different EHP generation mechanisms.

The value of n defines how far away from an ideal (no recombination in the depletion region). If $n = 1$ the diode is ideal, and if $n = 2$ recombination in the junction is very high, indicating there are many traps in the junction.

In reverse bias, on the other hand, generation current in the depletion region can add to the reverse bias current. These new EHPs add to generation current since the electric field separates them. Since the mechanism is trap-assisted generation, the electric field can play a role in how many carriers are generated using the trap state. The relationship between generation rate and electric field is square root dependence since the depletion width is proportional to the square root of the applied voltage. It should be noted that the traps in the semiconductor can be due to metal contamination in the fabrication process or crystal defects. Figure 4.26 shows the band diagram with trap states.

4.5.4 Ohmic Losses

In the ideal diode, we assumed the entire voltage is dropped across the depletion region and nothing outside. This assumption is fine for most cases; however, if the doping is high, ohmic loss in the neutral region can become significant enough to play a role. These losses can be modeled as resistance in the n and p regions and reduce the overall current.

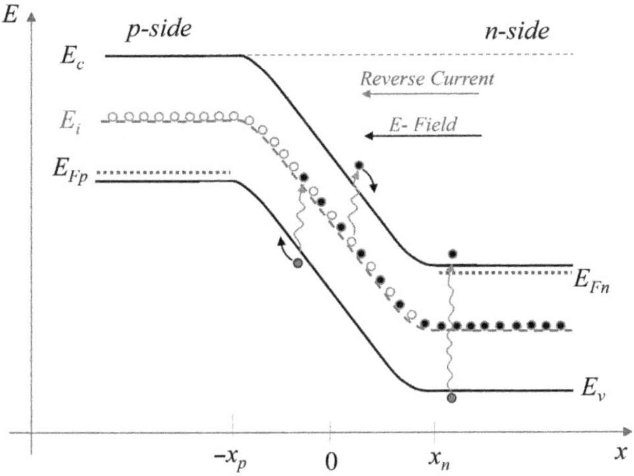

Figure 4.26 Band diagram showing the effect of trap states in reverse bias.

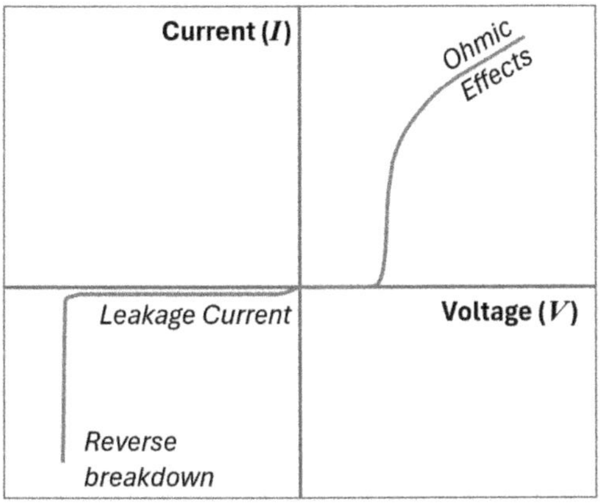

Figure 4.27 Diode *IV* characteristics with nonideal effects, showing reverse breakdown, leakage current, and ohmic losses.

Combining all the nonideal effects and plotting them side by side shows the effects as shown in Figure 4.27.

4.6 Chapter Summary

1) When *p*-type and *n*-type semiconductor regions are connected, electrons and holes diffuse due to concentration gradients. This diffusion leaves behind fixed

ionized dopant atoms, creating a depletion region and a built-in electric field that opposes further diffusion.

2) At equilibrium, the drift of carriers due to the built-in electric field balances the diffusion of carriers caused by concentration gradients. As a result, no net current flows, and a built-in potential is established across the junction.
3) Using the depletion approximation and Poisson's equation, the charge density, electric field, and potential across the *pn* junction are determined. The electric field exists only in the depletion region, reaches a maximum at the junction, and the depletion width depends on the doping concentrations on both sides.
4) Under reverse bias, the applied voltage increases the built-in potential and widens the depletion region, reducing majority-carrier diffusion. Under forward bias, the applied voltage reduces the built-in potential and narrows the depletion region, allowing majority carriers to diffuse across the junction and resulting in an exponential increase in current.
5) Real *pn* junctions exhibit nonideal behavior such as reverse bias breakdown, recombination and generation in the depletion region, turn-on voltage effects, and ohmic losses. These effects modify the ideal current–voltage characteristics of the diode.

End-of-Chapter Problems

Short Questions

P4.1 In a *pn* junction ____________________ is formed.

P4.2 A *pn* junction reaches __________ equilibrium, where diffusion and drift currents balance each other, resulting in zero net current.

P4.3 The total charge on *p*-side and *n*-side must be ______ to maintain charge neutrality.

P4.4 The peak electric field occurs at the ________ of the *pn* junction.

P4.5 The depletion region in the *pn* junction extends _________ in the lightly doped side.

P4.6 *pn* junction shows an _____________ *IV* curve.

P4.7 In reverse bias, the *pn* junction current is due to the________ of minority carriers in the depletion region.

P4.8 In forward bias, the *pn* junction current is due to ________ of majority carriers.

P4.9 Carrier ________________ in the junction can reduce the forward bias current or increase the reverse bias current and this considered a nonideal effect.

P4.10 Reverse bias breakdown voltage in a *pn* junction is due to __________ and _________.

Multiple Choice Questions

P4.11 Which of the following correctly describes the relationship between intrinsic carrier concentrations in silicon?
A $n = p = n_i$
B $n_p = n_i^2$
C Both (A) and (B)
D Neither (A) nor (B)

P4.12 At room temperature, intrinsic semiconductors conduct because:
A All electrons move freely in the conduction band.
B Covalent bonds are completely broken.
C Thermal energy generates EHPs.
D They are doped with both donor and acceptor impurities.

P4.13 Which statement about drift and diffusion currents in semiconductors is correct?
A Drift current is due to carrier motion from high to low concentration.
B Diffusion current is caused by an applied electric field.
C Drift and diffusion currents always oppose each other.
D Drift is due to electric field; diffusion is due to concentration gradient.

P4.14 The built-in voltage V_0 of a *pn* junction at equilibrium:
A Is typically 0.6–0.9 V for silicon at room temperature.
B Appears directly across the external terminals in open circuit.

C Decreases with increasing temperature.
D Is independent of doping concentration.

Numerical Problems

P4.15 Calculate the hole concentration in an n-type Si with concentration of donors equal to 10^{15} cm^{-3}? (Assume room temperature).

P4.16 Calculate the electron concentration in a p-type Ge sample where the acceptor concentration $N_a = 5 \times 10^{16}$ cm^{-3}. Assume $T = 300$ K.

P4.17 Calculate the intrinsic carrier concentration of Germanium at $T = 300$ K and $T = 600$ K. Assume that for Ge, $B = 1.66 \times 10^{15}$ and $E_g = 0.68$ eV is constant with temperature.

P4.18 Given an undoped Si material with bandgap $E_g = 1.12$ eV, material constant $B = 7.3 \times 10^{15}$ cm^{-3} K$^{-3/2}$ and Boltzmann's constant $k = 8.62 \times 10^{-5}$ eV/K

a) Calculate intrinsic carrier concentration of Si at $T = 600$ K.
b) What is the concentration of electrons and holes in intrinsic Si?

P4.19 A n-type piece of silicon with a length of 10 cm and a cross-sectional area of 0.05 cm × 0.05 cm is connected to 2 V potential difference. If the electron doping level is $n = 10^{17}$ cm^{-3}, calculate the total current flowing through the silicon bar at $T = 300$ K. (Assume $\mu_n = 1200$ cm^2/V·s and $\mu_p = 500$ cm^2/V·s, $T = 300$ K, $n_i = 1.5 \times 10^{10}$cm^{-3}, $E_g = 1.12$ eV).

P4.20 For a Si bar doped p-type at 10^{18} cm^{-3}, and sitting at temperature of $T = 400$ K, calculate:

a) Intrinsic concentration.
b) Concentration of electrons.
c) Conductivity of the bar.

P4.21 Assume room temperature. Draw and label (E_g, E_f, E_i, E_c, and E_v) the energy band diagram of Ge if it is doped with 10^{17} cm^{-3} acceptors.

P4.22 A Si sample is doped with 10^{16} donor atoms/cm^3. Find the hole concentration at equilibrium, p_0, at 400 K? Where is E_F relative to E_i?

P4.23 A Si sample is doped with 10^{16} cm^{-3} Boron and a certain number of donor atoms (N_d). The Fermi level is 0.36 eV above E_i at 300 K. Calculate value of N_d.

P4.24 Light is shining on Ge sample that is doped with acceptors of concentration 10^{17} cm^{-3}. This results in uniform optical generation of 10^{20} EHPs/cm^3 at $t=0$. Assume hole and electron lifetime $\tau_p=\tau_n=1$ μs. Calculate Fermi level before shining light and the quasi-Fermi levels after exposure to light.

P4.25 Consider a silicon solar cell where the concentration gradient of holes in the p-region is given by $dp/dx = -10^{21}$ cm^{-4}. The diffusion coefficient for holes in silicon, $D_p = 12$ cm^2/s. Calculate the hole diffusion current density J_p^{diff}.

P4.26 Si *pn* junction at $T=300$ K with $N_a = 5\times10^{16}$ cm^{-3} and $N_d = 2\times10^{16}$ cm^{-3}.
a) Find E_F–E_i on *n*-side and E_i–E_F on *p*-side.
b) Compute V_0.
c) Sketch the band diagram with labels.

P4.27 Repeat question P4.26 for GaAs at 300 K with $N_a = 1\times10^{17}$ cm^{-3} and $N_d = 5\times10^{16}$ cm^{-3}, using n_i(GaAs, 300 K) ≈ 2×10^6 cm^{-3}. Compare to Si and explain.

P4.28 Design Si *pn* junction at 300 K with $N_d = 1\times10^{16}$ cm^{-3}. Choose N_a so that $V_0 \approx 0.9$ V.

P4.29 For the Si junction with $N_a = 5\times10^{16}$ cm^{-3} and $N_d = 2\times10^{16}$ cm^{-3}, compute V_0 at $T=350$ K using a relationship for temperature-dependent n_i and assuming that bandgap is same as at 300 K.

P4.30 Quantify % change in V_0 from 300 to 400 K and estimate dV_0/dT numerically.

P4.31 A silicon diode conducts $I=1$ mA at 300 K. Estimate the change in forward voltage at 400 K ($n \approx 1$), assuming I_s follows $ni^2(T)$.

P4.32 Si *pn* junction at 300 K with $N_a = 1\times10^{16}$ cm^{-3} and $N_d = 4\times10^{17}$ cm^{-3}. Compute total depletion width W, partitions x_p and x_n, and peak electric field $|E_{max}|$ at zero bias.

P4.33 In problem P4.32, apply a reverse bias of 5 V and recompute W and $|E_{max}|$. Comment on which side widens more.

P4.34 Si diode at 300 K with $N_a = 1\times10^{17}$ cm^{-3}, $N_d = 5\times10^{16}$ cm^{-3}, $\mu_n = 1350$, $\mu_p = 480$ cm^2/V·s, $\tau_n = 2$ μs, $\tau_p = 0.5$ μs, and area = 100 μm × 100 μm. Compute I_s, then I at $V=0.65$ V and at $V=-5$ V. (Assume ideal diode.)

P4.35 A *pn* photodiode generates a light-current $I_L = 1$ mA at 300 K. Estimate open-circuit voltage V_{oc}.

P4.36 A *p*-type silicon sample is doped with $N_a = 1 \times 10^{16}$ cm^{-3}. Given $n_i = 1.5 \times 10^{10}$ cm^{-3}, mobilities $\mu_n = 1110$ cm^2/V · s, and $\mu_p = 400$ cm^2/V·s:

a) Calculate electron (n) and hole (p) concentrations at 300 K.
b) Compute the resistivity of the material.

P4.37 Explain what happens to the width of the depletion region in a *pn* junction when:

a) No bias is applied.
b) A reverse bias is applied.
c) A forward bias is applied.

Comment briefly on how these changes affect the junction current in each case.

P4.38 For a *pn* junction with $N_a = 1 \times 10^{18}$ cm^{-3}, $N_d = 1 \times 10^{16}$ cm^{-3}, and $n_i = 1.5 \times 10^{10}$ cm^{-3}, calculate

a) The built-in potential V_0 at 300 K.
b) The depletion width W.
c) Comment on the effect of increasing reverse bias on W.

Reference

1 Sze, S.M., Li, Y., and Ng, K.K. (2021). Physics of Semiconductor Devices, 4th ed. Wiley.

Reading Materials

1 Streetman, B.G. and Banerjee, S. (2015). Solid State Electronic Devices, globale. Prentice Hall.
2 Neamen, D. (2011). Semiconductor Physics and Devices: Basic Principles, 4th ed. McGraw-Hill.

5

Physics of Solar Cells

5.1 About the Chapter

The fundamental physics of solar cells is that photons with an energy larger than the bandgap of the semiconductor can be absorbed and converted to electron–hole pairs (EHP). If collected, these EHP can provide power to connected electrical devices much like a battery. However, these newly generated EHPs will recombine shortly after, rendering this process meaningless since the charge carriers are not collected so will not be used. The built-in electric field from the *pn* junction can be used to separate the carriers from each other before they recombine. This is the reason *pn* junctions are used to make solar cells. In this chapter, the physics of *pn* junction solar cells is explained. Also, the key solar cell parameters are derived including the open-circuit voltage, short-circuit current, efficiency, and fill factor (FF).

Key Topics

- *pn* junction operation under illumination (photodiode to solar cell)
- Photogeneration, carrier separation, and collection
- Open-circuit voltage and short-circuit current
- Maximum power point and fill factor
- Solar cell efficiency and bandgap limitations

5.2 Photodiode Physics

We have seen in Chapter 4 that the reverse-bias current in a *pn* junction originates from carrier generation within the depletion region and from carriers generated close enough to the junction to be collected, typically within a diffusion length.

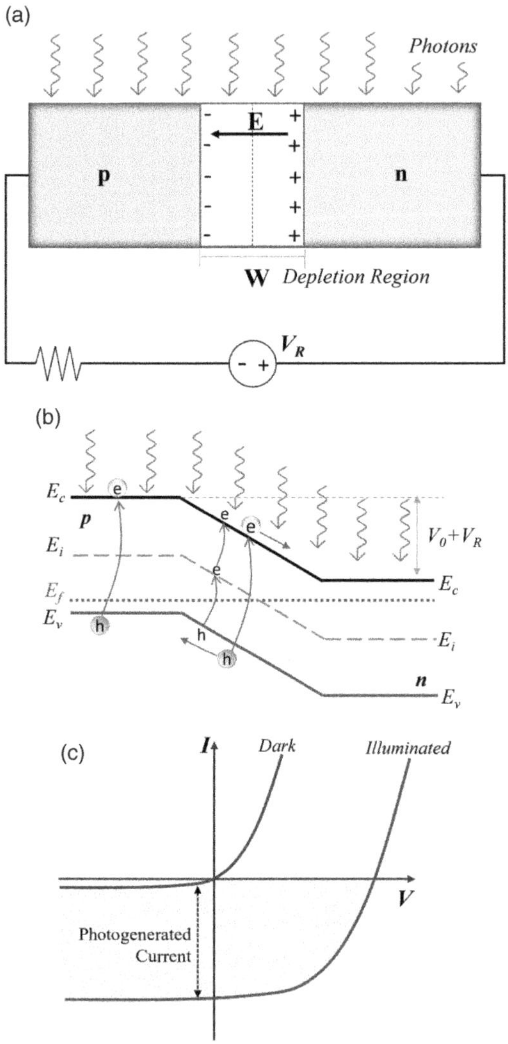

Figure 5.1 (a) Cross-section in reverse bias, (b) band diagram, and (c) *IV* curve of a *pn* junction under illumination.

Under illumination, absorbed photons create additional EHP, which increases the reverse-bias current. It is important to note that only photons absorbed within a diffusion length of the junction, or directly within the depletion region, contribute significantly to the reverse saturation current. Figure 5.1a shows a cross-section of a reverse-biased *pn* junction with incident photons being absorbed in the device. Figure 5.1b shows the corresponding energy band diagram under illumination, highlighting carrier generation and separation by the built-in electric field.

Under reverse bias, the depletion region supports a strong electric field and a potential drop of $V_o + V_R$. When photons are absorbed, EHPs are generated throughout the structure. EHPs created within the depletion region are immediately separated by the electric field: electrons are swept toward the n-side and holes toward the p-side. Carriers generated outside the depletion region can still contribute if they are close enough to the junction to diffuse into the depletion region before recombining. Figure 5.1c shows the effect of sunlight on the *IV* curve, where the illuminated curve shifts downward due to the increased reverse saturation current.

Let us calculate how many carriers are generated by defining three distinct regions that contribute to carrier collection: depletion region of width (W), the electron diffusion length (L_n) in the p-region, and the hole diffusion length (L_p) into the n-region. The diffusion length describes how far photogenerated carriers can travel by diffusion before recombining. This is visualized in Figure 5.2, which shows a 3D model of the *pn* junction under photon absorption.

The EHP generation rate due to photon absorption can be evaluated in each of the three regions: the p-region, the depletion region of width W, and the n-region. Let us define the amount of *EHPs* generated per second by photon absorption as g_{op} with units $cm^{-3}s^{-1}$. This definition applies to all photons with energy greater than or equal to the bandgap energy. The total photogenerated current is obtained by multiplying the number of collected carriers per second by the elementary

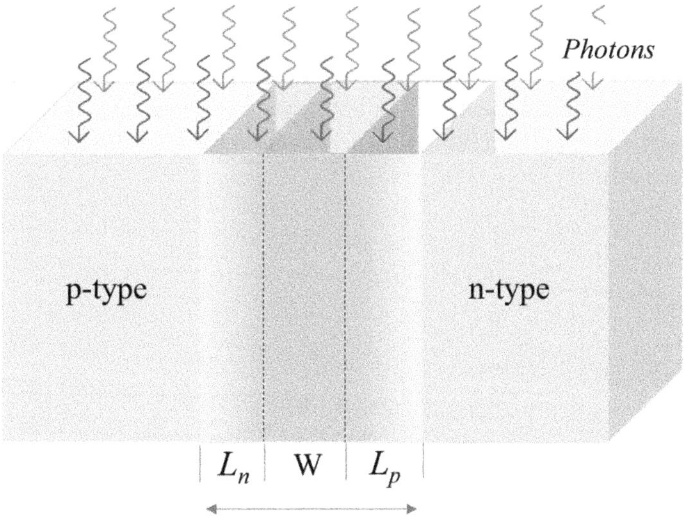

Figure 5.2 3D model of photon absorption regions (L_n, W, and L_p) in a *pn* junction.

charge q. Therefore, we can define the current due to photon absorption across the area A by I_p given as

$$I_p = qAg_{op}(W + L_n + L_p) \tag{5.1}$$

The higher the g_{op}, the higher the photogenerated current due to photon absorption. In addition, the wider depletion region and longer diffusion lengths allows for more photogenerated carriers to be collected, resulting in a higher current.

Example 5.1 ***Calculate the optical generation current***
Take a *Si pn* junction at 300 K, doped $N_d = 5 \times 10^{18}\ \text{cm}^{-3}$, $N_a = 1 \times 10^{16}\ \text{cm}^{-3}$. The junction has a square-shaped area, with side 0.5 cm. Junction has the following properties:

n-side:

$\mu_n = 100\ \text{cm}^2/\text{V} \cdot \text{s}$ and $\mu_p = 50\ \text{cm}^2/\text{V} \cdot \text{s}$
$\tau_n = 20$ ns and $\tau_p = 5$ ns

p-side:

$\mu_n = 1350\ \text{cm}^2/\text{V} \cdot \text{s}$ and $\mu_p = 450\ \text{cm}^2/\text{V} \cdot \text{s}$
$\tau_n = 1\ \mu\text{s}$ and $\tau_p = 0.5\ \mu\text{s}$

Assume a reverse-bias voltage of $V_r = 3$ V and $g_{op} = 3 \times 10^{19}\ \text{cm}^{-3}\ \text{s}^{-1}$. Calculate the optically generated current I_p.

Solution
The first step is to find the diffusion lengths on both sides. On the p-side, we need the electron diffusion length and, on the n-side, we need the hole diffusion length.

On the p-side electrons:

$$D_n = \mu_n \frac{kT}{q}$$

$$D_n = 1350 \times 0.0259$$

$$D_n = 34.97\ \text{cm}^2/\text{s}$$

On the n-side holes:

$$D_p = \mu_p \frac{kT}{q}$$

$$D_p = 50 \times 0.0259$$

$$D_p = 1.3\ \text{cm}^2/\text{s}$$

Diffusion length of electrons diffusing to the p-side is given by

$$L_n = \sqrt{D_n \tau_n}$$

$$L_n = \sqrt{34.97\,(1 \times 10^{-6})}$$

$$L_n = 59.14\ \mu\text{m}$$

Diffusion length of holes diffusing to the n-side

$$L_p = \sqrt{D_p \tau_p}$$

$$L_p = \sqrt{1.3(\,5 \times 10^{-9})}$$

$$L_p = 0.086\ \mu\text{m}$$

Next step is to find the depletion width using Eq. (4.37)

$$W = \sqrt{\frac{2\epsilon(V_0 + V_r)}{q}\left(\frac{1}{N_a} + \frac{1}{N_d}\right)} \tag{4.37}$$

Where V_0 is built-in potential,

$$V_0 = \frac{kT}{q} ln \frac{N_d N_a}{n_i^2}$$

$$V_0 = 0.0259 ln \frac{(5 \times 10^{18})(1 \times 10^{16})}{(1.5 \times 10^{10})^2}$$

$$V_0 = 0.856\ \text{V}$$

Plugging in the values and constants into Eq. (4.36) and using values: $\varepsilon_{r,Si} = 11.8$, $\epsilon_o = 8.85 \times 10^{-14}$ F/cm and $q = 1.6 \times 10^{-19}$ C.

$$W = \sqrt{\frac{2(11.8)(8.85 \times 10^{-14})(0.858 + 3)}{1.6 \times 10^{-19}}\left(\frac{1}{1 \times 10^{16}} + \frac{1}{5 \times 10^{18}}\right)}$$

$$W = 7.1 \times 10^{-5}\ \text{cm}$$

The total optical current is:

$$I_p = qAg_{op}(W + L_n + L_p)$$

$$I_p = 1.6 \times 10^{-19} \times 0.5 \times 0.5 \times 3 \times 10^{19}\,(7.1 \times 10^{-5} + 5.91 \times 10^{-3} + 8.6 \times 10^{-6})$$

$$I_p = 7.19\ \text{mA}$$

So, the current density is

$$J_p = I_p/A = \frac{7.19\ \text{mA}}{0.5\ \text{cm}^2} = 28.8\ \text{mA/cm}^2$$

We can add Eq. (5.1) to Eq. (4.40) to define an equation for the total current. The diode equation with no photon absorption (dark) from Eq. (4.40) is $I_d = I_0(e^{qV_a/kT} - 1)$. The total current ($I_t$) is given by

$$I_t = I_d + I_p \tag{5.2}$$

$$I_t = I_0 \left(e^{qV_a/kT} - 1\right) + q \cdot A \cdot g_{op}(W + L_n + L_p) \tag{5.3}$$

If there is no photon absorption, then the g_{op} will be zero and total current will be the diode current in the dark.

5.2.1 Operation of Photodiode as a Solar Cell

The application of photodiode depends on which *IV* quadrant the device operates in. In quadrants 1 and 3, the voltage and current are positive, yielding positive power ($P > 0$). In this operation, the diode can act as a photodetector. The absorbed photons change the current, and this is detectable. In quadrant 4, the voltage is positive and current is negative, yielding a negative power. This is the operation for a solar cell, since power is *delivered* to the circuit and the *pn* junction acts like a battery. This is illustrated in Figure 5.3.

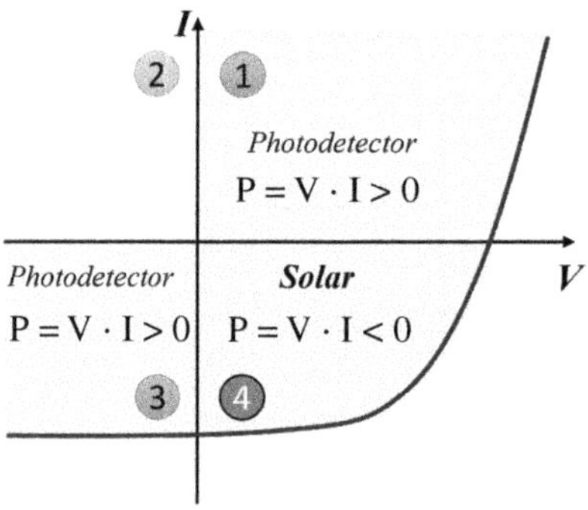

Figure 5.3 *IV* characteristics of *pn* junction under illumination, indicating which quadrant the diode operates as photodetector or solar cell.

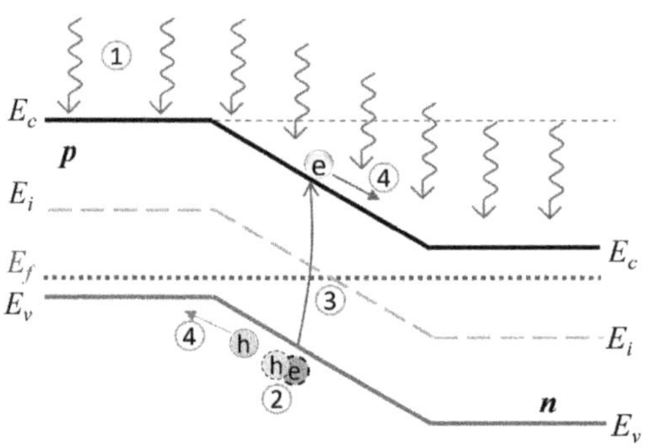

Figure 5.4 Solar cell operation steps: (1) photon absorption, (2) EHP generation, (3) charge separation, and (4) drift under reverse bias.

Figure 5.4 is the band diagram of the *pn* junction in reverse bias that shows the solar cell operation steps: (1) photon absorption, (2) EHP generation, (3) charge separation, and (4) drift.

5.3 Open-Circuit Voltage and Short-Circuit Current

One of the key features of solar cells is the appearance of a voltage when the current is zero and a current when the voltage is zero. This is clearly shown in the plot of Figure 5.3. We can define these two values as V_{oc} (open-circuit voltage) and J_{sc} (short-circuit current). For convenience, the *IV* curve in Figure 5.3 is plotted with positive current values as in Figure 5.5. This is the standard convention to plot the solar *IV* curve.

Using Eq. (5.3), the total current can be written as

$$I_t = I_0 \left(e^{qV_a/kT} - 1\right) - I_p \tag{5.4}$$

Solving for the voltage when current is zero, we can determine open-circuit voltage, V_{oc}

$$0 = I_0(e^{qV_{oc}/kT} - 1) + I_p$$

We then solve for V_{oc} through the following algebraic steps:

$$I_0 \left(e^{qV_{oc}/kT} - 1\right) = I_p$$

$$e^{qV_{oc}/kT} = \frac{I_p}{I_0} + 1$$

$$\frac{qV_{oc}}{kT} = ln\left(\frac{I_p}{I_0} + 1\right)$$

$$V_{oc} = \frac{kT}{q} ln\left(\frac{I_p}{I_0} + 1\right) \tag{5.5}$$

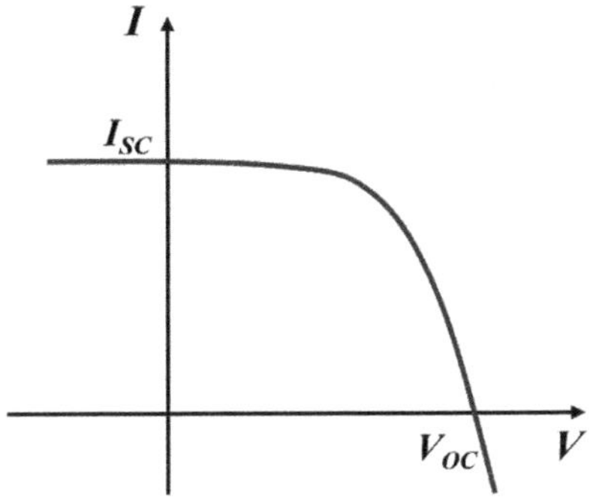

Figure 5.5 Standard solar cell *IV* curve with positive current convention.

This can be simplified when $\frac{I_p}{I_0} \gg 1$, leading to

$$V_{oc} \approx \frac{kT}{q} ln\left(\frac{I_p}{I_o}\right) \quad (5.6)$$

This formula highlights some important physics on how to increase the V_{oc}: increasing the amount of optically generated carriers (I_p) or decreasing the reverse saturation current in the dark (I_0).

The current density when the voltage is zero is the short-circuit current density

$$J_{sc}(V_a) = \frac{1}{A}(I_0\left(e^{qV_a/kT} - 1\right) + I_p) \quad (5.7)$$

When $V_a = 0$, then

$$J_{sc} = I_p/A \quad (5.8)$$

J_{sc} is simply the total current density achieved from optical absorption only. For a high-performance single-junction Si solar cell, J_{sc} can range between 35 and 45 mA/cm^2.

Example 5.2 ***Finding*** V_{oc}

Take a *GaAs pn* junction at 300 K doped $N_d = 1 \times 10^{18}$ cm^{-3}, $N_a = 5 \times 10^{17}$ cm^{-3}. The junction has a square area of 0.25 cm^2. Assume photogenerated current $I_p = 1.91$ mA.

n-side:

$D_n = 7.77$ cm^2/s and $D_p = 2.59$ cm^2/s
$\tau_n = 5$ ns and $\tau_p = 2$ ns

p-side:

$D_n = 51.8$ cm^2/s and $D_p = 5.18$ cm^2/V · s
$\tau_n = 10$ ns and $\tau_p = 5$ ns

Calculate the V_{oc}.

Solution

$$V_{oc} = \frac{kT}{q} ln\left(\frac{I_p}{I_0} + 1\right)$$

Find I_0

$$I_0 = qA\left(\frac{D_p}{L_p}p_n + \frac{D_n}{L_n}n_p\right)$$

Diffusion length on p-side

$$L_n = \sqrt{D_n \tau_n}$$

$$L_n = \sqrt{51.8\,(10 \times 10^{-9})}$$

$$L_n = 5.18 \times 10^{-3}\ \mu\text{m}$$

Diffusion length on n-side

$$L_p = \sqrt{D_p \tau_p}$$

$$L_p = \sqrt{5.18(\,5 \times 10^{-9})}$$

$$L_p = 1.61\ \mu\text{m}$$

Find p_n: minority carrier hole concentration on the n-side

$$np = n_i^2$$

Since $N_d \gg n_i$ $(1 \times 10^{18} \gg 2 \times 10^6)$, then $n \approx N_d$

$$p_n = \frac{n_i^2}{N_d}$$

$$p_n = \frac{(2 \times 10^6)^2}{1 \times 10^{18}}$$

$$p_n = 4 \times 10^{-6}\ \text{cm}^{-3}$$

Find n_p: minority carrier electron concentration on the p-side

$$np = n_i^2$$

Since $N_a \gg n_i$ $(5 \times 10^{17} \gg 2 \times 10^6)$, then $p \approx N_a$

$$n_p = \frac{n_i^2}{N_a}$$

$$n_p = \frac{(2 \times 10^6)^2}{5 \times 10^{17}}$$

$$n_p = 8 \times 10^{-6}\ \text{cm}^{-3}$$

Substitute values to find I_0

$$I_0 = qA\left(\frac{D_p}{L_p}p_n + \frac{D_n}{L_n}n_p\right)$$

$$I_0 = -1.6 \times 10^{-19}(0.25)^2 \left(\frac{2.59}{1.6 \times 10^{-4}}(4 \times 10^{-6}) + \frac{51.8}{5.18 \times 10^{-7}}(8 \times 10^{-6}) \right)$$

$$I_0 = 8 \times 10^{-18}\ \text{A}$$

Finally, we obtain

$$V_{oc} = 0.0259\, ln \left(\frac{1.91 \times 10^{-3}}{8 \times 10^{-18}} + 1 \right)$$

$$V_{oc} = 0.857\ \text{V}$$

5.4 Power Max and Fill Factor

The maximum power that the solar cell can provide is equal to the maximum area under the *IV* curve. This is true since power is equal to current multiplied by voltage ($P = IV$). The maximum rectangle can be found by finding the inflection point in the curve. Let us define, V_m and I_m points from the graph, which gives the maximum power rectangle as shown in Figure 5.6.

Using Eq. (5.4), the power delivered by a solar cell can be calculated as

$$P(V) = I \cdot V = V\left(I_0\left(e^{qV_a/kT} - 1\right) - I_p\right) \tag{5.9}$$

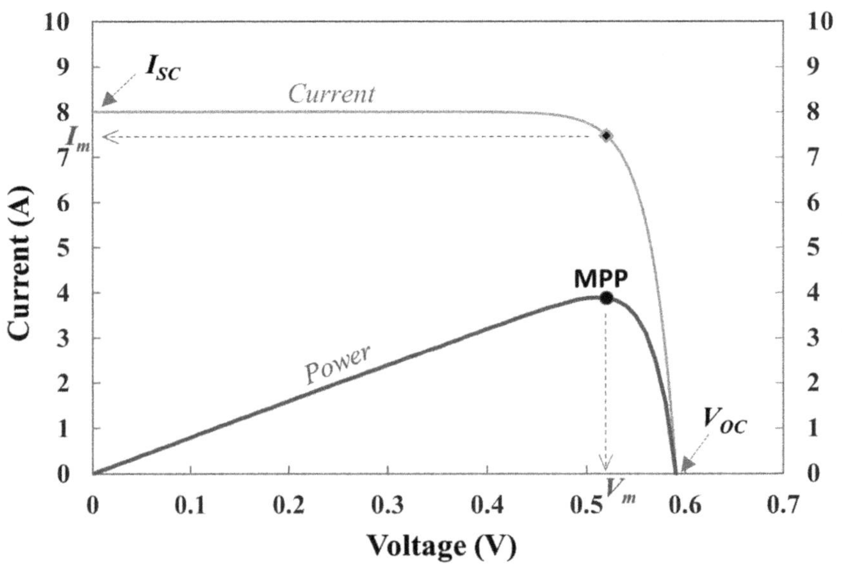

Figure 5.6 Solar cell *IV* characteristics with current (A) on left and power (W) on the right axis, indicating points of I_{SC}, V_{OC}, I_m, V_m, and the maximum power point (MPP).

Equation (5.9) describes power as a function of voltage in the solar cell. To find the maximum power value, we find the voltage value at which the derivative, $\frac{dP}{dV} = 0$.

$$\frac{dP}{dV} = V \cdot (I_0(e^{qV_a/kT} - 1)) - (VI_p) = 0$$

Separating into three terms, we get

$$\frac{dP}{dV} = VI_0\left(e^{qV_a/kT}\right) - VI_0 - VI_p = 0$$

Using the chain rule and derivatives of exponential, we get the following expression:

$$\frac{dP}{dV} = V_m I_0\left(e^{qV_m/kT}\right) - I_o - I_p = 0$$

$$\frac{dP}{dV} = I_0\left(e^{qV_m/kT}\right)\left(1 + V_m\frac{q}{kT}\right) - I_0 - I_p = 0$$

$$\left(e^{qV_m/kT}\right)\left(1 + V_m\frac{q}{kT}\right) = \frac{I_0 + I_p}{I_0} \tag{5.10}$$

Here V_m indicates voltage with maximum power delivered by a solar cell.

Using this formula, the value of V_m can be solved by trial and error (iteration) or numerical methods.

Example 5.3 ***Finding the maximum voltage and current***

A Si *pn* junction solar cell at $T = 300$ K has $I_0 = 1 \times 10^{-11}$ A. The generated photocurrent in the cell is $I_p = 150$ mA. Determine the maximum voltage (V_m, the maximum current (I_m), and the maximum power (p_{max}).

Solution

Using Eq. (5.10),

$$\left(e^{qV_m/kT}\right)\left(1 + V_m\frac{q}{kT}\right) = \frac{I_0 + I_p}{I_0}$$

Substituting known quantities and $\frac{q}{kT} \approx 0.029$, we get

$$\left(e^{38.6\ V_m}\right)(1 + 38.6\ V_m) = \frac{1 \times 10^{-11} + 150 \times 10^{-3}}{1 \times 10^{-11}}$$

$$\left(e^{38.6\ V_m}\right)(1 + 38.6\ V_m) = 1.5 \times 10^{10}$$

Trial 1: $V_m = 0.5\ \text{V} \rightarrow 4.89 \times 10^9 \neq 1.5 \times 10^{10}$
Trial 2: $V_m = 0.55\ \text{V} \rightarrow 3.69 \times 10^{10} \neq 1.5 \times 10^{10}$
Trial 3: $V_m = 0.53\ \text{V} \rightarrow 1.65 \times 10^{10} \neq 1.5 \times 10^{10}$
Trial 4: $V_m = 0.5277\ \text{V} \rightarrow 1.50 \times 10^{10} = 1.5 \times 10^{10}$

With V_m found, I_m can be found from (Eq. (5.2)) (Total Current at V_m) using the given I_p

$$I_m = I_0(e^{qV_m/kT} - 1) - I_p$$

$$I_m = 1 \times 10^{-11}(e^{38.6 \cdot 0.5277} - 1) - 150 \times 10^{-3}$$

$$I_m = 143\ \text{mA}$$

Finally

$$P_{\max} = V_m I_m = (0.5277)(143)\ \text{mW}$$

Figure 5.7 shows the example of solar cell *IV* characteristics, with maximum rectangle and the larger rectangle shaded.

From Figure 5.7, the area of the larger rectangle, A_1 is $J_{sc}V_{oc}$ and the smaller, A_2 is J_mV_m. Based on this, we can define a new term called the *fill factor* (FF), which is the area of the smaller rectangle divided by the area of the larger rectangle. So, the FF is defined as

$$FF = \frac{J_m V_m}{J_{sc} V_{oc}} \tag{5.11}$$

and since

$$P_{\max} = J_m V_m \tag{5.12}$$

then,

$$P_{\max} = FF \cdot J_{sc} \cdot V_{oc} \tag{5.13}$$

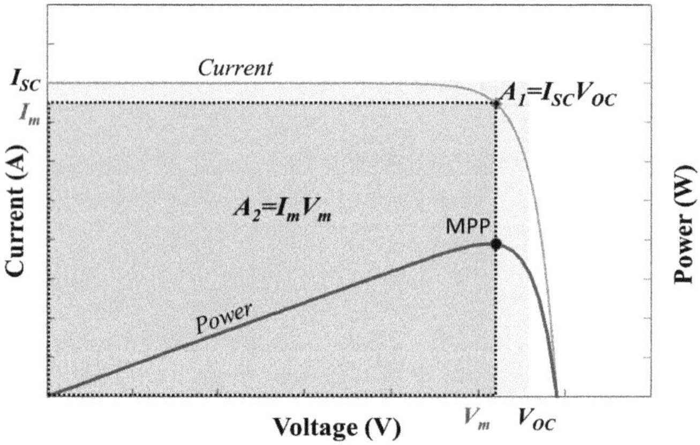

Figure 5.7 Solar *IV* curve with maximum rectangle, and the larger rectangle labeled.

Physically, the FF reflects how closely the *IV* curve of the solar cell resembles an ideal rectangle. A higher FF indicates lower internal resistive losses and better diode quality, contributing to more efficient energy conversion. In an ideal solar cell with no resistive losses and perfect diode behavior, the FF would be 1. However, in practical devices, typical values range from 0.70 to 0.85 for high-efficiency silicon cells.

Example 5.4 ***Fill factor***

For the cell described in Example 5.3, calculate the FF.

Solution

First step is to find V_{OC}

$$V_{oc} = \frac{kT}{q} ln\left(\frac{I_p}{I_0} + 1\right)$$

$$V_{oc} = 0.0259\, ln\left(\frac{150 \times 10^{-3}}{1 \times 10^{-11}} + 1\right)$$

$$V_{oc} = 0.607\ \text{V}$$

To calculate *FF*, we use Eq. (5.11), and the fact that $J = IA$, we get

$$FF = \frac{J_m V_m}{J_{sc} V_{oc}} = \frac{I_m V_m}{I_{sc} V_{oc}}$$

We found in Example 5.3 that $V_m I_m = 75.46$ mW

We can also assume that $I_{sc} = I_p = 150$ mA (from example 5.3)

$$FF = \frac{75.46\ (\text{mW})}{150\ (\text{mA}) \cdot 0.607(\text{V})}$$

$$FF = 82.9\%$$

5.5 Efficiency

The next important performance factor for solar cells is efficiency. Efficiency is often cited and reported as the key performance factor. Solar cell efficiency, also known as power conversion efficiency (PCE), represents the ratio of the electrical power output at the MPP on the *IV* curve divided by the input power from the sun. It reflects the performance of the solar cell itself and depends on various factors such as the spectrum and intensity of the incident sunlight and the temperature of the solar cells. These design factors will be considered in detail in Chapter 6. Let us define efficiency for solar with symbol (η) as

$$\eta = \frac{P_{\max}}{P_{in}} \tag{5.14}$$

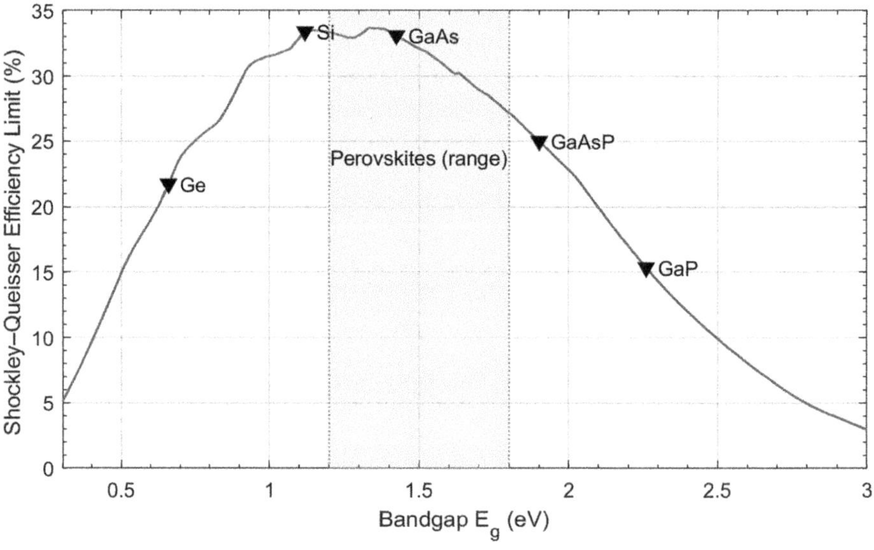

Figure 5.8 Efficiency vs. bandgap (Shockley–Queisser limit).

and substituting known parameters,

$$\eta = \frac{FF \cdot J_{sc} \cdot V_{oc}}{P_{in}} \tag{5.15}$$

P_{in} is the total power from the sun that was explained and calculated in Chapter 3. The AM 1.5 global spectrum has an integrated power of 1000 W/m^2. Equation (5.15) shows that to increase efficiency, we must increase the FF, V_{oc} or J_{sc}. It is important to note that the bandgap (E_g) is the fundamental limitation and main physical parameter that determines the maximum efficiency possible. The bandgap (E_g) will limit the current and the voltage that can be achieved. The higher the bandgap the higher the voltage but less optically generated current. As the bandgap decreases, the voltage produced by the solar cell decreases, while the optically generated current increases. As such, an optimum bandgap exists to give the maximum possible efficiency. Figure 5.8 plots efficiency of solar cell vs. bandgap of the material, highlighting the trade-off between current and voltage. The plot shows that the best bandgap is 1.4 eV with Shockley–Queisser limit for efficiency of a single-junction solar cell around 33%. It should be noted that Si, which dominates the solar market, has a bandgap of 1.1 eV.

Example 5.5 ***Efficiency of solar cell***

A Si solar *pn* junction solar cell has a V_{oc} of 0.65 V, J_{sc} = 37 mA/cm^2 and a fill factor FF = 85%. Find the efficiency of the solar cell under $AM1.5G$ spectrum.

Solution

Using Eq. (5.15), for P_{out}/P_{in}, and for $AM1.5G$ total input power, we have

$$P_{in} = 1000\ \text{W/m}^2 = 0.1\ \text{W/cm}^2$$

$$\eta = \frac{FF \cdot J_{sc} \cdot V_{oc}}{P_{in}}$$

Changing the current to A/cm^2 and FF to a decimal.

$$\eta = \frac{(0.85)(3.7 \times 10^{-2})(0.65)}{0.1}$$

$$\eta = 0.2044 \text{ or } 20.44\%$$

At this point, we have covered enough physics to talk about efficiency in a practical way. Recall from Chapter 2 that the solar spectrum tells us how many photons arrive at the device, while the QE tells us how many of those photons are collected and contribute toward photogenerated current density (J_{ph}). Hence, QE is such a powerful "bridge quantity" between optics and electrical performance.

In Chapter 2, we defined external quantum efficiency EQE(λ), as the fraction of incident photons that produce collected electrons (Eq. (2.3)). We also introduced internal quantum efficiency IQE(λ), to separate optical losses from electrical collection losses (Eq. (2.4)). Finally, we showed that the photogenerated current density is found by weighting the incident photon flux density by EQE(λ) and integrating across wavelength (Eq. (2.5)). In this chapter, we saw that photogenerated current density is equivalent to short-circuit current density ($J_{ph} = J_{SC}$).

Example 5.6 ***Find the current and IQE***

Suppose at $\lambda = 600$ nm your solar cell has EQE = 0.85. Under *AM* 1.5*G*, a typical photon flux is about 5.7×10^{14} photons cm^{-2} s^{-1} nm^{-1}.

a) Calculate a current density generated over a 10-nm wavelength interval.
b) Calculate IQE, if transmission $\mathcal{T} = 0$ and $R = 0.12$ over a 10-nm wavelength "slice."

Solution

Given $\Delta\lambda = 10$ nm, $\Phi = 5.7 \times 10^{14}$ photons cm^{-2} s^{-1} nm^{-1}, and EQE = 0.85.

a) $J \approx q\,\Phi\,\text{EQE} \cdot \Delta\lambda = (1.602 \times 10^{-19}) \times (5.7 \times 10^{14}) \times 0.85 \times 10 \approx 0.78\ \text{mA/cm}^2$.

Note that summing "slices" like this across the above-bandgap range of frequencies gives the total J_{SC}.

b) $\text{IQE} \approx \text{EQE}/(1 - R) = 0.85/(1 - 0.12) = 0.966$.

Note: Here, we do not integrate because flux was assumed constant across 10-nm wavelength interval.

The reason we bring the QE topic back now is simple. In Chapter 2, we could describe how an *ARC* reduces optical reflection and helps more photons enter the device. However, at that stage, we intentionally did not discuss what happens after absorption, because carrier collection depends on recombination physics that we are discussing now. In this chapter, we finally have the tools to connect optics to carrier transport, and this connection is essential for understanding how efficiency improves through an increase in J_{SC}.

Now recall that an *ARC* mainly improves optics by reducing the reflectance spectrum $R(\lambda)$, which increases the number of photons that enter the solar cell. However, reducing reflection is only beneficial if those additional photons can be converted into collected current. Furthermore, a single-layer *ARC* minimizes reflection around one design wavelength, while reflection can increase at other wavelengths. This is why *ARC* optimization should not be thought of as minimizing reflection where the sunlight is strongest, but rather minimizing reflection where the device can efficiently convert absorbed photons into collected carriers.

This point becomes especially important once we consider carrier lifetime and diffusion length. Photons absorbed far from the junction generate EHP that must survive long enough to reach the depletion region (or the collecting contact) before recombining. Therefore, even if a wavelength corresponds to a large photon flux $\Phi(\lambda)$, it may still contribute weakly to J_{SC} if carrier collection is poor at that depth. On the other hand, reducing reflection in a wavelength region where collection is strong, meaning where IQE(λ) is high, it can produce a larger gain in short-circuit current and solar cell efficiency. For example, thin-film devices often benefit from *ARC* designs that favor shorter wavelengths, since long-wavelength photons can penetrate deeper and generate carriers beyond the effective diffusion length, reducing the probability of collection.

Next, we discuss an example that illustrates an important design rule: when choosing where to minimize reflection, the most impactful wavelength is not determined by $\Phi(\lambda)$ alone, but by the product $\Phi(\lambda) \cdot \text{IQE}(\lambda)$, since only photons that are both abundant and well collected contribute strongly to J_{sc}.

Example 5.7 ***Find the current and IQE***

A solar cell is evaluated at two different wavelengths for *ARC* design. The goal is to choose a design wavelength that produces the larger gain in J_{sc} when reflection is reduced.

Assume the device is sufficiently thick so transmission is negligible. The *ARC* is assumed to only change reflection, not recombination, so IQE(λ) remains unchanged.

Two wavelength options are considered:

Option A: design at $\lambda_1 = 500$ nm
Option B: design at $\lambda_2 = 850$ nm

The photon flux density values under AM 1.5G illumination are approximated as:

Φ (500 nm) = 2.0×10^{15} photons cm^{-2} s^{-1} nm^{-1}
Φ (850 nm) = 1.2×10^{15} photons cm^{-2} s^{-1} nm^{-1}

Before adding ARC, the reflectance is measured to be

$$R(500\text{ nm}) = 0.30 \text{ and } R(850\text{ nm}) = 0.20.$$

From device measurements, the IQE is: IQE(500 nm) = 0.55 and IQE(850 nm) = 0.92

Assume a narrow wavelength band $\Delta\lambda = 20$ nm around each design wavelength. An ARC is used and it reduces reflectance at the design wavelength by $\Delta R = 0.10$

a) For each wavelength option, estimate the increase in EQE caused by the ARC.
b) For each wavelength option, estimate the increase in current density ΔJ_{sc} contributed by that wavelength band.
c) Which wavelength is the better ARC design target, and why?

Solution

For negligible transmission, we recall from Eq. (2.4) that: IQE(λ) ≈ EQE(λ)/[1 – $R(\lambda)$]

So: EQE(λ) ≈ IQE(λ)[1 – $R(\lambda)$]

Because the ARC only changes $R(\lambda)$, the change in EQE is:

$$\Delta\text{EQE}(\lambda) = \text{IQE}(\lambda)\Delta(1 - R) = \text{IQE}(\lambda)(\Delta R)$$

a) Increase in EQE

At 500 nm:

$$\Delta\text{EQE}(500) = \text{IQE}(500)\Delta R = (0.55)(0.10) = 0.055$$

At 850 nm:

$$\Delta\text{EQE}(850) = \text{IQE}(850)\Delta R = (0.92)(0.10) = 0.092$$

b) Increase in current density

Using Eq. (2.5), $\Delta J_{sc} \approx q\,\Phi(\lambda)\,\Delta\text{EQE}(\lambda)\,\Delta\lambda$

At 500 nm:

$\Delta J_{sc}(500) = q(2.0 \times 10^{15})(0.055)(20) = \Delta J_{sc}(500) = (1.602 \times 10^{-19})(2.2 \times 10^{15})$
$\Delta J_{sc}(500) = 3.52 \times 10^{-4}$ A cm^{-2} = 0.352 mA cm^{-2}

At 850 nm:

ΔJ_{sc} (850) = $q(1.2 \times 10^{15})(0.092)(20) = (1.602 \times 10^{-19})(2.208 \times 10^{15})$
$\Delta J_{sc} = 3.54 \times 10^{-4}$A cm^{-2} = 0.354 mA cm^{-2}

c) Best ARC target and interpretation

The two options produce nearly the same current gain:

$$\Delta J_{sc}(500) \approx 0.352 \text{ mA cm}^{-2}$$

$$\Delta J_{sc}(850) \approx 0.354 \text{ mA cm}^{-2}$$

Even though the photon flux is higher at 500 nm, the device collects carriers much more efficiently at 850 nm because IQE is significantly higher. As a result, reducing reflection near 850 nm can be just as effective, or even slightly better, than reducing reflection where more photons arrive.

5.5.1 Practical Considerations

Scientists and researchers designing solar cells are often interested in both EQE and IQE, as they provide information about unique solar cell design features. EQE combines optics and electrons, while IQE isolates the electronics part only. So if IQE is high but EQE is low, we should investigate improving optics, such as ARC, texturing, improving layer thickness, and refractive-index matching. If, on the other hand, both EQE and IQE are low in some region, we need to improve carrier collection, often by passivation, better junction design, using materials with better mobility/lifetime, etc. In short, QE is the wavelength-by-wavelength "report card" of a solar cell, pinpointing optical and electrical losses as a function of wavelength.

5.6 Chapter Summary

1) When a *pn* junction absorbs sunlight, EHPs are created. The built-in electric field separates the carriers, so they do not recombine, and the carriers created near or in the junction are added to the reverse-bias current. Solar cells operate in quadrant 4 of the *IV* curve, where negative power is delivered to the external circuit.
2) One of the key features of solar cells is the appearance of a voltage when the current is zero and a current when the voltage is zero. The voltage when the current is zero is the open-circuit voltage, and the current when the voltage is zero is the short-circuit current. These are the two main parameters that determine the performance of a solar cell.
3) The maximum power that the solar cell can provide is equal to the maximum area under the *IV* curve. The MPP is defined by the voltage and current at which this area is maximized. The FF is the area of the smaller rectangle divided by the area of the larger rectangle and reflects how closely the *IV* curve resembles an ideal rectangle.

4) Solar cell efficiency represents the ratio of the electrical power output at the MPP divided by the input energy from the sun. The efficiency depends on the open-circuit voltage, short-circuit current, and FF. The efficiency is limited by the bandgap of the material used, which determines the tradeoff between current and voltage.
5) QE is a practical tool for identifying which wavelengths contribute most to J_{SC} and efficiency. Since J_{SC} is directly proportional to the spectral product $\Phi(\lambda)\,\mathrm{EQE}(\lambda)$, ARC optimization should not focus only on minimizing reflection where the solar photon flux is highest, but where improved optical coupling will produce the largest current gain, typically in wavelength ranges where $\Phi(\lambda)\,\mathrm{IQE}(\lambda)$ is large. This becomes especially important in thin-film devices, where limited thickness and shorter diffusion lengths can reduce collection of carriers generated deeper in the absorber, particularly at longer wavelengths.

End-of-Chapter Problems

Multiple Choice Questions and Answers

P5.1 The open-circuit voltage occurs when:

A $V=0$

B $J=0$

C $P=0$

D $J=J_0$

P5.2 The short-circuit current density occurs when:

A $V=0$

B $J=0$

C $V=V_{OC}$

D $P=P_{\max}$

P5.3 *FF* is defined as:

A $FF=\dfrac{J_{SC}}{V_{OC}}$

B $FF=\dfrac{J_mV_m}{J_{SC}V_{OC}}$

C $FF=\dfrac{J_0}{J_{ph}}$

D $FF=\dfrac{P_{in}}{P_{\max}}$

P5.4 Efficiency is defined as:

A $\eta = \frac{V_{OC}}{J_{SC}}$

B $\eta = \frac{P_{in}}{P_{\max}}$

C $\eta = \frac{P_{\max}}{P_{in}}$

D $\eta = \frac{J_0}{P_{in}}$

P5.5 A *pn* junction is illuminated with constant g_{op}. If $(L_n + W + L_p)$ is doubled while the area A stays the same, the optical current I_{ph} becomes:

A Half

B Same

C Doubled

D Quadrupled

P5.6 Given that $g_{op} = 1 \times 10^{19}$ cm^{-3} s^{-1}, $L_n + W + L_p = 10$ μm $= 1 \times 10^{-3}$ cm, and $A = 0.01$ cm^2. Photogenerated current I_{ph} is approximately

A 1.6 μA

B 16 μA

C 0.16 mA

D 1.6 mA

P5.7 A solar cell has $V_{OC} = 0.70$ V, $J_{SC} = 30$ mA/cm^2, $V_m = 0.55$ V, and $J_m = 27$ mA/cm^2. Find *FF*.

A 0.52

B 0.71

C 0.89

D 1.10

P5.8 If V_{OC}, J_{SC}, and *FF* all increase while P_{in} is unchanged, then efficiency η will:

A Decrease

B Stay the same

C Increase

D Become independent of P_{in}

P5.9 A solar cell has $V_{OC} = 0.65$ V, $J_{SC} = 35$ mA/cm^2, and $FF = 0.80$, $P_{in} = 100$ mW/cm^2.

Approximate efficiency.
A 9.1%
B 18.2%
C 26.0%
D 33%

P5.10 At a certain wavelength, a measured cell has: IQE = 0.95 and EQE = 0.60. Dominant limitation is most likely:
A Electrical collection loss
B Optical loss (reflection/transmission)
C Bandgap too small
D *FF* loss

Fill in the Blanks

P5.11 Solar cell operation requires carrier ________________ before recombination occurs.

P5.12 The depletion region has a strong ________________ field that sweeps carriers.

P5.13 The voltage at zero current is called the ________________ voltage.

P5.14 The current density at zero voltage is called the ________________ current density.

P5.15 The Shockley–Queisser limit for Si single-junction solar cell is approximately ___ %.

Numerical Problems

P5.16 A GaAs *pn* junction ($E_g = 1.42\,\text{eV}$) at 300 K has $N_d = 2 \times 10^{18}\,\text{cm}^{-3}$, $N_a = 1 \times 10^{16}\,\text{cm}^{-3}$. The device area is a square with 0.2-cm-long side. Assume: $\mu_n = 8500\,\text{cm}^2/(\text{V·s})$, $\mu_p = 400\,\text{cm}^2/(\text{V·s})$, $\tau_n = 10\,\text{ns}$, and $\tau_p = 5\,\text{ns}$. For $V_r = 2\,\text{V}$ and $g_{op} = 2 \times 10^{19}\,\text{cm}^{-3}\,\text{s}^{-1}$, calculate the optically generated current.

P5.17 Explain why only carriers generated within a diffusion length of the junction contribute to current. What happens to carriers generated outside this region?

P5.18 A Si solar cell with area 0.5 cm^2 generates $J_{ph} = 35$ mA/cm^2 under illumination. The diode reverse saturation current density is $J_o = 2 \times 10^{-12}$ A/cm^2. Find V_{oc} at 300 K.

P5.19 A material has high carrier mobility but short lifetime. Another has long lifetime but low mobility. Which would you choose for solar cells, and why? Support with formula references.

P5.20 Two wavelengths are candidates for ARC optimization. At those wavelengths, a solar cell has the following IQE values and input irradiance:

- Case 1: $\Phi_1 = 2 \times 10^{15}$, $\text{IQE}_1 = 0.4$
- Case 2: $\Phi_2 = 1 \times 10^{15}$, $\text{IQE}_2 = 0.9$

Assume, same ΔR and $\Delta \lambda$ and that transmission is negligible. Which of the cases will gives larger improvement in current, ΔJ_{SC}?

A Case 1 only
B Case 2 only
C Equal
D Cannot be determined

P5.21 Suppose you observe the following on the measurements for a solar cell:

- 350–420 nm: EQE is low and IQE is also low (computed with $R = 0.03$, $\mathcal{T} \approx 0$).
- 550 nm: R is high (10%) and IQE = 0.95.
- 800–900 nm: strong long-wavelength roll-off.

Tasks:

a) Identify the dominant loss mechanism in the UV.
b) If you reduce R at 550 nm from 10 to 2% ($\mathcal{T} \approx 0$ and IQE = 0.95), what is the new EQE?
c) Give two design fixes for the long-wavelength roll-off.

P5.22 A *pn* junction has the following parameters:

$$g_{op} = 2 \times 10^{19}\ \text{cm}^{-3}\ \text{s}^{-1}$$

$W = 2\ \mu\text{m}$, $L_n = 8\ \mu\text{m}$, $L_p = 4\ \mu\text{m}$
Square area with side 0.2 cm
Calculate the optically generated current I_{ph}.

P5.23 A solar cell ($T = 300$ K) has the following properties: $J_{ph} = 35$ mA/cm^2 and $J_0 = 2 \times 10^{-12}$ A/cm^2. Find V_{OC}.

P5.24 A solar cell has $J_{ph} = 28$ mA/cm^2 and active area $A = 2$ cm^2. Find J_{SC} and total short-circuit current I_{SC}.

P5.25 A thin-film CdTe solar cell generates 28 mA/cm^2 under AM 1.5 illumination. Calculate the short-circuit current density, J_{sc}, and total current for device area = 2 cm^2.

P5.26 A solar cell has $V_{OC} = 0.65$ V, $J_{SC} = 35$ mA/cm^2, and maximum power values: $V_m = 0.52$ V, $J_m = 34$ mA/cm^2. Find FF.

P5.27 A solar cell has the following parameters: $V_{OC} = 0.65$ V, $J_{SC} = 35$ mA/cm^2, and $FF = 0.85$. For input power $P_{in} = 100$ mW/cm^2(AM1.5G), calculate the efficiency of this cell.

P5.28 A solar cell has $J_{sc} = 35$ mA/cm^2, $V_{oc} = 0.65$ V, and $P_{max} = 18$ mW/cm^2.
a) Calculate FF.
b) If series resistance R_s increases, qualitatively sketch how the IV curve and FF change.

P5.29 A GaAs solar cell ($E_g = 1.42$ eV) has $J_{sc} = 30$ mA/cm^2, $V_{oc} = 1.0$ V, and $FF = 0.85$. For AM 1.5 input power 100 mW/cm^2, calculate efficiency. Compare with Shockley–Queisser limit (~33%).

P5.30 At $\lambda = 600$ nm, a solar cell has EQE = 0.5 and $\Phi(\lambda) = 5.7 \times 10^{14}$ photons cm^{-2} s^{-1} nm^{-1}. Calculate the current density contribution J over that wavelength slice $\Delta\lambda = 10$ nm.

P5.31 At $\lambda = 650$ nm, EQE = 0.7 and $\Phi = 6 \times 10^{14}$ cm^{-2} s^{-1} nm^{-1}. For the wavelength slice $\Delta\lambda = 20$ nm, estimate photogenerated current J.

P5.32 Two wavelength options are considered for improving reflectance by $\Delta R = 0.10$, assuming $\mathcal{T} \approx 0$.
- At 500 nm: $\Phi = 2.0 \times 10^{15}$, IQE = 0.4
- At 850 nm: $\Phi = 1.2 \times 10^{15}$, IQE = 0.8

$$\Delta\lambda = 20 \text{ nm}$$

Find ΔJ_{SC} for each band and state which is the better ARC target.

P5.33 A cell of active area $A = 0.25\,\text{cm}^2$ is illuminated at $\lambda = 550\,\text{nm}$ with a narrow band $\Delta\lambda = 5\,\text{nm}$. The photon flux density is $\Phi = 6.0 \times 10^{14}\,\text{cm}^{-2}\,\text{s}^{-1}\,\text{nm}^{-1}$. The measured short-circuit current is $I_{SC}(\lambda) = 20\,\mu\text{A}$. Reflectance $R = 0.06$, transmission $\mathcal{T} = 0$.

Tasks:

a) Compute EQE(λ).
b) Compute IQE(λ).
c) Report the current density contribution over the 5-nm slice (wavelength interval).

6

Solar Cell Design

6.1 About the Chapter

After understanding the physics of solar cells and the key solar cell parameters, it is important to understand the key design considerations to maximize efficiency. While it is true that a solar cell is simply a *pn* junction, many additional features and design factors are needed to make a solar cell with the highest possible efficiency. This chapter describes the main parts of an individual solar cell that include the absorber layer, emitter, anti-reflective coating (ARC), and the metal finger contacts. Moreover, the physics of the design is discussed for maximum performance with typical dimensions.

Key Topics

- Solar cell structure and layer roles
- Absorber layer design and thickness trade-offs
- Emitter design, doping, and carrier collection
- Optical absorption and reflection losses
- Anti-reflection coating design principles
- Metal finger contacts, resistance, and shading
- Advanced and emerging solar cell designs

6.2 Solar Cell Cross Section

As described in Chapter 5, a solar cell is a *pn* junction designed to absorb photons, generate electron–hole pairs, separate them before recombining, and finally collect the carriers. A typical cross-section for a complete solar cell is shown in Figure 6.1 with the various layers highlighted.

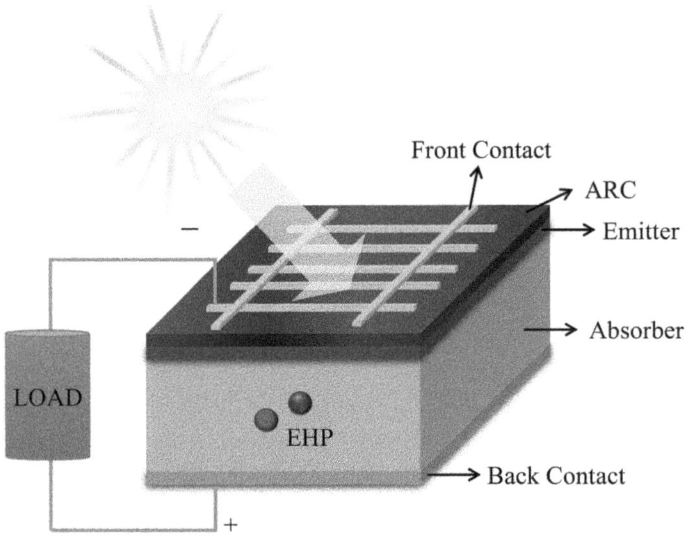

Figure 6.1 Illustration of typical solar cell structure.

Each layer in the solar cell plays an important role in achieving the highest efficiency. The *pn* junction part has two main regions: the absorber and emitter. The solar cell is designed to create and separate electron–hole pairs in absorber layer, where the separated electrons are swept into the emitter. This is accomplished by using the built-in electric field. Typically, the absorber is *p*-type and emitter is *n*-type for maximum performance, since the electron mobility is higher than hole mobility, allowing for higher carrier collection efficiency.

In addition, this configuration aligns with physics motion of generated charge carriers due to the sunlight absorption and minimizes recombination losses. The electrons, which are minority carriers in the *p*-type material, diffuse toward the *n*-type emitter. This movement aligns with built-in electric field at the *pn* junction, which helps to separate the carriers efficiently. This effect is shown by band diagram in Figure 6.2.

Since electrons have higher mobility in *n*-type and longer diffusion lengths than holes in *p*-type, they are more likely to reach the junction and be collected before recombining. In contrast, if the structure were reversed with an *n*-type absorber and *p*-type emitter, the minority carriers would be holes with less mobility and inherently provide less current. The slower mobility increases the probability they recombine before reaching the emitter, reducing the efficiency of the solar cell. Therefore, using a *p*-type absorber and *n*-type emitter not only supports better carrier collection but also enhances the overall device performance.

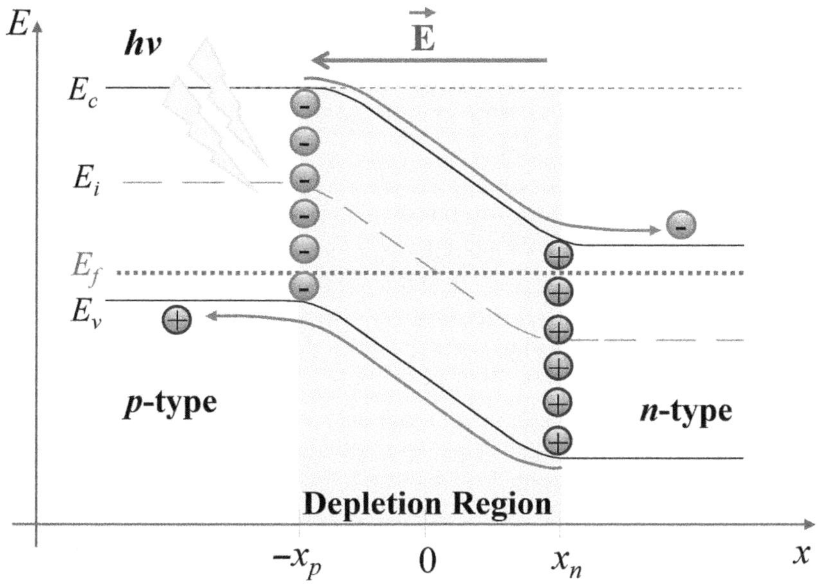

Figure 6.2 Band diagram illustration of the movement of electrons and holes in a *pn* junction under illumination.

6.3 Absorber Layer

The absorber layer is heart of the solar cell since it is the location where most of the photons are absorbed, and majority of the electron–hole pairs are created. It is the most important layer in the solar cell design, as it determines how many carriers are generated and how much power the solar cell can produce. For maximum performance, key properties that need to be considered are the bandgap, absorption coefficient, and thickness. For choosing the bandgap, remember that only photons with energy equal to or more than the bandgap are absorbed. As discussed before, a tradeoff between open-circuit voltage (V_{oc}) and short-circuit current (I_{sc}) for maximum efficiency leads to an optimum bandgap. It turns out the optimum bandgap for solar cell absorber layer is around 1.4 eV, as shown in Figure 5.8.

Besides the bandgap, thickness of the absorber layer is a key and critical parameter. This depends on the absorption coefficient, as discussed in Chapter 2. Ideally, it is best to use materials that have a very high absorption coefficient, so less material is needed. Silicon has a high absorption coefficient in the ultraviolet and visible regions (300–600 nm), where light is absorbed within just a few microns. However, in the near-infrared region (600–800 nm), the absorption coefficient decreases, requiring tens to hundreds of microns for effective absorption.

Beyond 800 nm, absorption becomes even weaker, making it necessary for significantly thicker layers. As a result, conventional silicon solar cells typically use a thickness of around 150–200 μm. This ensures that most of the solar spectrum, including the less energetic near-infrared photons, is absorbed effectively, balancing performance with material cost.

As described in Chapter 3, optical absorption in a semiconductor is governed by the Beer–Lambert law, which describes the exponential decay of optical intensity as it gets absorbed deeper into the material. Let us recall that from Chapter 2, when light of initial intensity, I_0, enters a semiconductor, the intensity at depth x is given by $I(x) = I_0 e^{-\alpha x}$, where α is the absorption coefficient (Eq. (2.10)). While α quantifies how strongly a material absorbs light, it is often more intuitive to consider the corresponding absorption length, defined as $L = 1/\alpha$ (Eq. (2.9)). The absorption length represents a characteristic depth scale over which significant absorption occurs. In other words, at a depth equal to L, the optical intensity has dropped to $1/e$ of its initial incidence value, meaning that approximately 63% of the incident light has been absorbed at one absorption depth. Plotting absorption length as a function of wavelength provides direct insight into how far photons travel within the material before being absorbed, which is particularly relevant for solar cell thickness and carrier collection considerations. Figure 6.3 presents the absorption length $L = 1/\alpha$ as a function of wavelength for several common photovoltaic absorber materials.

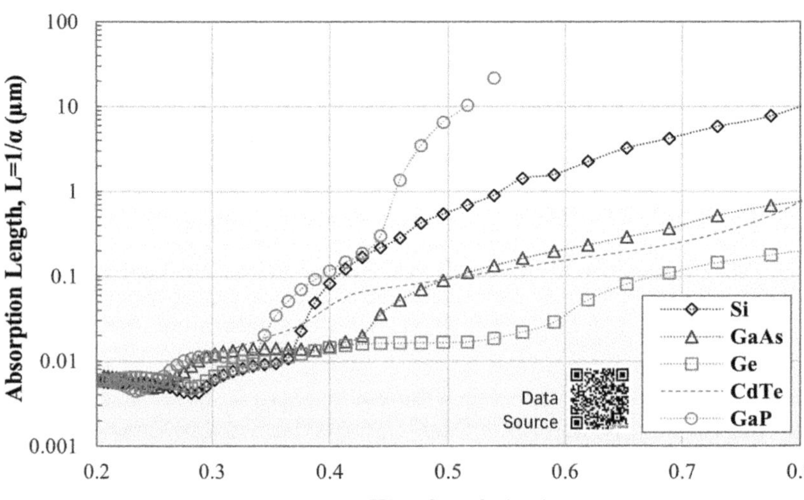

Figure 6.3 Absorption length $L = 1/\alpha$ (μm) as a function of wavelength (μm) for Si, GaAs, Ge, CdTe, and GaP materials. *Source:* Adapted from [1].

From Figure 6.3, it can be seen that for all materials, approximately 63% of the incident light at short wavelengths in the ultraviolet range is absorbed within roughly the first 100 nm of material. However, this behavior changes significantly at longer wavelengths. For example, germanium (Ge) can absorb about 63% of incident light at wavelengths up to approximately 0.5 μm within less than 20 nm of material, whereas silicon (Si) requires a thickness on the order of 600–700 nm to absorb the same fraction of light at this wavelength. The strong absorption in Ge is primarily due to its narrow bandgap.

If GaAs or CdTe are compared to Si, it is evident that both materials also require significantly thinner absorber layers to achieve the same level of absorption, which is a consequence of their direct bandgap. Direct-bandgap semiconductors such as GaAs and CdTe, therefore, exhibit short absorption lengths across most of the visible spectrum, typically on the order of a fraction of a micrometer to a few micrometers, indicating that approximately 63% of the entering light is absorbed within this shallow depth. In contrast, indirect-bandgap silicon exhibits much longer absorption lengths, particularly at longer wavelengths near its band edge, where the absorption length can reach tens of micrometers or more. As a result, photons at these wavelengths travel further into silicon before being absorbed, explaining the need for relatively thick wafers or optical light-trapping schemes in silicon solar cells. Materials such as gallium phosphide (GaP), which have a wider bandgap, display a rapidly increasing absorption length beyond their absorption edge, reflecting their limited absorption of longer-wavelength photons.

Example 6.1 ***Calculate Si absorber layer thickness***

You are designing a silicon solar cell and want to ensure 90% absorption of incident light at two different wavelengths: 400 and 800 nm. Calculate the required thickness of the silicon absorber layer for each wavelength.

Solution

First step is to get the absorption coefficient for Si at 400 and 800 nm. We can get it from the chart in Figure 6.3.

From the graph at 400 nm, $\alpha = 1 \times 10^5\ \text{cm}^{-1}$
From the graph, at 800 nm, $\alpha \approx 0.1\ \mu\text{m}^{-1}$
Converting to cm^{-1}: $\alpha \approx 0.1\ \mu\text{m}^{-1} = 0.1 \times 10^4 = 1 \times 10^3\ \text{cm}^{-1}$

The fraction of absorbed light as a function of depth x is given by the Beer–Lambert law:

$$A = 1 - e^{-\alpha x}$$

To find the thickness required for 90% absorption ($A = 0.9$), we solve for x:

$$x = -\frac{ln(1 - A)}{\alpha}$$

Absorption of 400 nm wavelength:

$$x = -\frac{ln(1 - 0.9)}{1 \times 10^5} = 0.2303\ \mu m$$

Solve for x with 90% absorption and for 800 nm

$$x = -\frac{ln(1 - 0.9)}{1 \times 10^3} = 2.30 \times 10^{-3}\ cm = 23\ \mu m$$

To absorb 90% of 400 nm light, silicon requires a thickness of approximately 0.23 μm.

To absorb 90% of 800 nm light, silicon requires a thickness of approximately 23 μm.

This large increase in required thickness at longer wavelengths reflects the much weaker absorption of silicon near its band edge and explains the need for relatively thick wafers and optical light-trapping schemes in silicon solar cells.

State-of-the-art silicon solar cells typically use absorber layers with a thickness ranging from 100 to 200 micrometers, optimized for light absorption, carrier collection, and material cost. This thickness is sufficient to absorb over 90% of the usable solar spectrum.

GaAs, on the other hand, is a highly efficient semiconductor material used in advanced solar cells and optoelectronic devices. It has a direct bandgap and a very high absorption coefficient, typically exceeding 10^4 to 10^5 cm^{-1} across most of the solar spectrum. This means *GaAs* can absorb sunlight within a very short distance, often less than 1 μm, for visible wavelengths. Even in the near-infrared region, GaAs maintains strong absorption, allowing for effective photon capture with minimal thickness. As a result, GaAs solar cells can achieve high efficiency with absorber layers as thin as 2–3 μm, significantly thinner than traditional silicon-based cells. This property makes GaAs ideal for lightweight, high-performance applications such as space photovoltaics and multi-junction solar cells.

It should be noted, high-efficiency Si solar cells use a lightly doped base region (typically *p*-type with doping concentrations around 10^{15} to 10^{16} cm^{-3}) to minimize the recombination in the base and maximize the collection.

Example 6.2 ***Calculate GaAs absorber layer thickness***

Repeat Example 6.1 for GaAs absorber layer. Compare results to Si case in example 6.1.

Solution

First step is to get the absorption coefficient for Si at 400 and 800 nm. We can get it from the chart in Figure 6.3.

From the graph in Figure 6.3, we get:

At 400 nm, $L \sim 0.015$ μm so $\alpha \approx \frac{1}{0.015\ \mu m} \times \frac{1\ \mu m}{10^{-4}\ cm} \approx 6.67 \times 10^5\ cm^{-1}$
At 800 nm, $L \sim 0.7$ μm, so $\alpha \approx \frac{1}{0.7\ \mu m} \times 10^4\ cm^{-1} \approx 1.43 \times 10^4\ cm^{-1}$
Use Beer–Lambert law to calculate the absorption:

$$A = 1 - e^{-\alpha x}$$

Solve for x with 90% absorption and for 400 nm

$$x = -\frac{ln(1 - A)}{\alpha}$$

$$x = -\frac{ln(1 - 0.9)}{6.67 \times 10^5} = 3.45 \times 10^{-6}\ cm = 0.0345\ \mu m$$

Solve for x with 90% absorption and for 800 nm

$$x = -\frac{ln(1 - 0.9)}{1.43 \times 10^4} = 1.61 \times 10^{-4}\ cm = 1.61\ \mu m$$

GaAs thickness for 90% absorption at 400 nm: 0.035 μm (about 35 nm)
GaAs thickness for 90% absorption at 800 nm: 1.61 μm

Compared to silicon, GaAs requires a much thinner absorber to achieve 90% absorption at the same wavelengths. This is because GaAs is a direct-bandgap semiconductor with a much larger absorption coefficient, so photons are absorbed within a much shorter depth.

6.4 Emitter

To design the emitter, the key factors are thickness and doping. The main role of the emitter is to form the *pn* junction and create the built-in field, which is needed to separate the charge carriers. So theoretically, a very thin layer is only needed to achieve this. For example, if the absorber is *p*-type with concentration $1 \times 10^{16}\ cm^{-3}$, then the *n*-type emitter can be doped higher. For the best design, the depletion region and electric field should be mostly in the *p*-type absorber region for maximum collection. This is achieved with a heavily doped *n*-type emitter. Typical doping for the *n*-type emitter ranges from 1×10^{18} to $1 \times 10^{20}\ cm^{-3}$.

In terms of thickness, two things need to be considered: diffusion of electrons and photon absorption. Electrons must diffuse from the absorber layer to the emitter before recombining. Since electrons pass through the emitter/absorber interface, it is very important to reduce interface defects with proper passivation. These defects can be the result of dangling bonds at interface or surface contamination. In addition, if the emitter is too thick, the electron must diffuse over a longer distance through the heavily doped region with low carrier lifetime, increasing the probability of recombination before being collected. Any loss of generated electron–hole pairs in the emitter is considered parasitic absorption of the light that would have otherwise passed to the absorber region and contributed to photogenerated current. The emitter thickness should ideally be comparable to or smaller than the electron diffusion length within the emitter region. This ensures efficient collection of minority carriers generated near the emitter–base junction. If the emitter is too thick, many of these generated electrons will recombine before reaching the junction, reducing device performance. On the other hand, if the emitter is too thin, the effect of interface defects will increase, so there should be an optimum thickness that provides a balance between these two extremes.

Ideally, all photons should be absorbed in the absorber, not in the emitter, because they have a higher chance of being collected if generated near the electric field. However, typically, photons with very high energy for short wavelengths between 300 and 500 nm will be mostly absorbed in the emitter. These photons will create electron–hole pairs in the emitter and can contribute to the overall photogenerated current, depending on the design and emitter material quality. If the emitter is very thick, generated electron–hole pairs may recombine before being collected. Also, if there is a large amount of interface defects at the emitter/absorber interface, they can also recombine before being collected. However, if those electron–hole pairs are collected, the high-energy photons can also contribute to the overall current.

Example 6.3 ***Estimate absorption in the emitter***

A silicon solar cell has a 20 nm thick emitter. Using the data from Figure 6.3, estimate the percentage of incident light at wavelengths 300 and 500 nm that is absorbed within the emitter. Assume single-pass absorption and neglect reflection.

Solution

Emitter thickness $t = 20\ \text{nm} = 0.02\ \mu\text{m}$

From Figure 6.3 (absorption length plot), read the silicon absorption length:

At 300 nm, $L = 0.08\ \mu\text{m}$, so $\alpha = \frac{1}{0.08\ \mu\text{m}} = 12.5\ \mu\text{m}^{-1}$

At 500 nm, $L = 0.6\ \mu\text{m}$, so $\alpha = \frac{1}{0.6\ \mu\text{m}} = 1.67\ \mu\text{m}^{-1}$

Using Beer–Lambert law to calculate the absorption in the emitter:

$$A = 1 - e^{-\alpha x}$$

At 300 nm wavelength, emitter absorbs $A_{300} = 1 - e^{-12.5 \times 0.02} = 0.22$ of incident light.

At 500 nm wavelength, emitter absorbs $A_{500} = 1 - e^{-1.67 \times 0.02} = 0.033$ of incident light.

So approximately 22% of incident light at 300 nm wavelength is absorbed in the emitter, while only ~3.3% of the incident light at 500 nm wavelength is absorbed in the emitter.

A typical thickness for emitter is around 0.2 – 0.5 μm. This thickness with high doping can allow enough absorption between 300 and 500 nm wavelength range and a strong electric field for separation. The other important factor in designing the emitter is the doping concentration. High conductivity is needed to allow the best possible carrier transport. In addition, a high electric field in the junction is key to getting the most carrier separation. These two factors can be accomplished by high doping concentrations.

Example 6.4 ***Is Emitter absorption desirable?***

Solution

In most solar-cell designs, significant absorption in the emitter is not desirable because carriers generated very close to the front surface are more likely to recombine due to higher surface recombination and because the emitter region is often heavily doped, which can reduce minority-carrier lifetime. In addition, the device is typically designed so that most photons are absorbed in the base (absorber), where collection is more efficient. For these reasons, strong emitter absorption is often treated as a parasitic optical loss, especially for short-wavelength light. One should keep in mind that absorption in the emitter is not always "wasted" if the emitter is well passivated and the junction field helps collect carriers, but as a general design rule, we try to minimize it.

6.5 Anti-Reflective Coatings

Another major challenge in solar cell design is the reflection of the incident light at the front surface. The concept of reflection was introduced in Chapter 2 in detail. In a silicon solar cell, photons initially travel through air and encounter the silicon surface. Due to the large refractive-index mismatch between air and silicon, a significant fraction of the incident light is reflected. The reflectance between Si and air at normal incidence can be estimated using the Fresnel expression for reflection using $n_{si} = 3.5$ at a specific wavelength from Eq. (2.14) is

$$R = \left(\frac{n_{si} - n_{air}}{n_{si} + n_{air}}\right)^2 = \left(\frac{3.5 - 1}{3.5 + 1}\right)^2 = 0.308 = 31\%$$

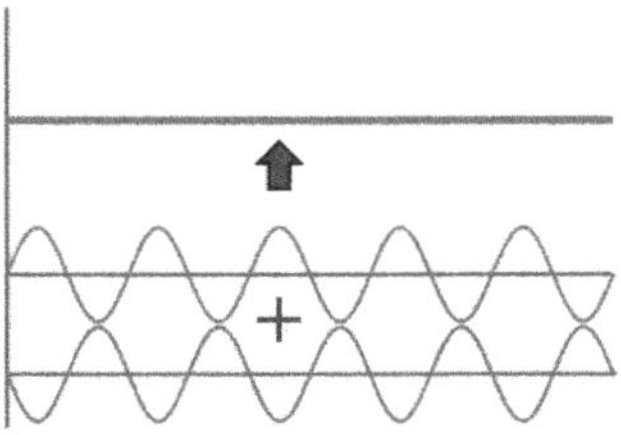

Figure 6.4 Illustration of how destructive interference results in zero reflectance.

This means that 31% of the photons are reflected and hence not absorbed in the solar cell. This reflected light is considered an optical loss and a waste of energy, as it reduces the number of photons available for absorption and carrier generation. The good news is that surface reflection can be reduced by adding a specific coating to the surface of the Si. These coatings, called anti-reflection coatings (ARC), are thin layers of materials like silicon nitride (SiN_x) or titanium dioxide (TiO_2) deposited on top of the solar cell surface. ARCs are designed so that the light reflected from the top surface of the coating and the light reflected from the coating–silicon interface undergo destructive interference. This means that the reflected waves are out of phase and partially cancel each other out. As a result, more photons are absorbed and transmitted to the absorber layer, increasing the current generated by photons and ultimately the efficiency. Figure 6.4 illustrates how destructive interference results in zero reflection for single-layer ARC design.

Example 6.5 ***Anti-reflection coating***

Assume a single-layer ARC is applied to a silicon solar cell.

a) Determine the required thickness of an Si_3N_4 coating to minimize reflection at a wavelength of 550 nm.
b) Repeat the calculation for a TiO_2 coating.

Solution

The refractive index of the various materials at a specific wavelength is assumed to be:

$$n_{air} = 1, \quad n_{\mathrm{Si_3N_4}} = 2, \quad n_{\mathrm{TiO_2}} = 2.4, \quad n_{\mathrm{Si}} = 3.5$$

For a single-layer ARC at normal incidence, destructive interference of the reflected waves occurs when the coating thickness d satisfies the quarter-wave condition:

$$d = \frac{\lambda}{4n_s}$$

a) For Si_3N_4 ARC

$$d = \frac{\lambda}{4n_{Si_3N_4}}$$

$$d = \frac{550}{4 \times 2} = 68.75\ \text{nm}$$

b) For TiO_2 ARC

$$d = \frac{\lambda}{4n_{TiO_2}}$$

$$d = \frac{550}{4 \times 2.4} = 57.29\ \text{nm}$$

Required Si_3N_4 coating thickness: ≈ 69 nm
Required TiO_2 coating thickness: ≈ 57 nm

As demonstrated in this example, both the refractive index of the coating material and its thickness play a critical role in determining the effectiveness of an ARC. Properly designed coatings can significantly reduce surface reflection at the target wavelength. To further suppress reflection over a broader spectral range, multi-layer ARC can be employed. By combining materials with different refractive indices and carefully optimizing their thicknesses, multi-layer ARCs enable lower overall reflectance across the solar spectrum than single-layer designs, as is discussed earlier in Chapter 2.

6.6 Finger Contacts

An additional challenge in solar cell optimization is designing the contacts in a way that limits the series resistance, but it also minimizes shading losses. As a thought experiment, imagine only a single contact is placed on the surface (Figure 6.5a). While this will allow most of the light to be absorbed in the solar cell, it will also create a large amount of resistance since all the electrons will flow to one location over a long distance. One solution to address this challenge is making finger metal contacts to spread the current to multiple contacts, while still allowing light to be absorbed in the solar cell (Figure 6.5b).

An illustration of a top view of a commercial solar cell with fingers is shown in Figure 6.6, showing the metal fingers.

This combination of fingers and busbar creates an optimum resistance vs. photon shading effect. Typical metal finger coverage in commercial solar cells ranges from 3 to 8% of the total cell area. Most commercial silicon solar

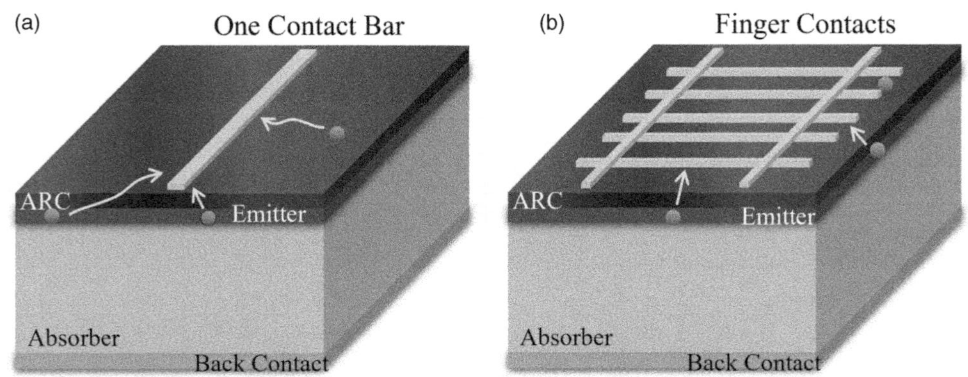

Figure 6.5 Illustration of the effect of contact fingers on collection of carriers: (a) example with only one contact bar showing longer paths for collection and (b) example with many contact fingers showing how collection path is shortened.

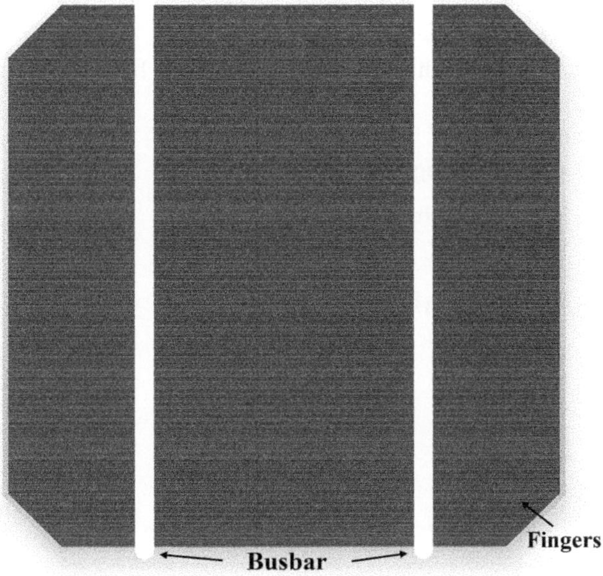

Figure 6.6 Illustration of a typical commercial solar cell contact structure, showing two large busbars and many thin, closely spaced fingers.

cells use screen-printed silver fingers with widths ranging from 55 to 80 μm. For screen-printed silicon solar cells, finger spacing typically ranges from 1 to 3 mm using Ag, Al, Cu, and Ni. Moreover, the transparent conductive oxides (TCO) are used to combine transparency and conductivity. TCOs can be used on top of solar cells with metal fingers and can reduce the overall resistance and improve carrier collection.

Example 6.6 ***Metal fingers***
A 50 mm × 50 mm square solar cell uses metal fingers that are 60 μm wide and spaced 2 mm apart (center-to-center).

a) Calculate the total number of fingers.
b) Calculate the total area covered by the metal fingers.
c) Determine the percentage of the cell area covered by the fingers.

Solution
a) Find the number of fingers, N:

$$N = \frac{\text{Width of Solar Cell}}{\text{Spacing Between}} = \frac{50\,\text{mm}}{2\,\text{mm}} = 25\ \text{fingers}$$

So, the total number of fingers is: $\boxed{N = 25}$

b) Area covered by metal fingers
First find the shaded area for 1 finger.
Each finger runs across the full width of the cell (50 mm) and has a width of 60 μm = 0.06 mm:
$A_{finger} = 0.06\ \text{mm} \times 50\ \text{mm} = 3.00\ \text{mm}^2$
The total shaded area due to all 25 fingers is:

$$A_{\text{total}} = 25 \times 3.00\ \text{mm}^2 = 75\ \text{mm}^2$$

c) Percentage of cell area shaded

$$A_{cell} = 50 \times 50\ \text{mm} = 2500\ \text{mm}^2$$

% shaded by metal:

$$\frac{75\ \text{mm}^2}{2500\ \ \text{mm}^2} = 3\%$$

Example 6.7 ***Metal fingers***
Repeat Example 6.6 if the finger width is 40 μm.

Solution
a) Find the number of fingers:

$$\frac{\text{Width of Solar Cell}}{\text{Spacing Between}} = \frac{50\ \text{mm}}{2\ \text{mm}} = 25\ \text{fingers}$$

b) Calculate the shaded area from metal fingers
First find the shaded area for 1 finger:
$A_{finger} = 0.04\ \text{mm} \times 50\ \text{mm} = 2.00\ \text{mm}^2$
Multiply by the total number of fingers.

$$A_{total} = 2.00\ \text{mm}^2 \times 25 = 50\ \text{mm}^2$$

c) Calculate % shaded for the total cell:

$$A_{cell} = 50\ \text{mm} \times 50\ \text{mm} = 2500\ \text{mm}^2$$

% shaded by metal:

$$\frac{50\ \text{mm}^2}{2500\ \text{mm}^2} = 2\%$$

As can be seen reducing the finger width can reduce the shading. There is an optimum width, spacing to balance shading and conductivity.

6.7 Novel Design Methods

Over the years, improvements to the solar cell design have been carried out to increase solar cell conversion efficiency. For the absorber layer, the key change has been to use novel materials that have higher absorption coefficient compared to Si. One material that has recently shown great promise is perovskites that use a hybrid organic–inorganic lead or tin halide as the light-absorbing layer. These materials follow the formula ABX_3, where A is a large cation (e.g. methylammonium), B is a metal cation (e.g. lead), and X is a halide anion (e.g. iodide), offering high efficiency and tunability for next-generation photovoltaics. Their absorption coefficient is extremely high in the ultraviolet and visible regions of the solar spectrum, often exceeding 10^5 cm^{-1}, which allows them to absorb light within just a few hundred nanometers. Even in the near-infrared range (700–900 nm), where absorption is weaker, perovskites still maintain reasonable performance with coefficients around 10^3 cm^{-1}, requiring only a few microns of thickness. As a result, a perovskite layer as thin as 500 nm to 1 μm is typically sufficient to absorb most of the solar spectrum. This ultrathin absorption capability enables lightweight, flexible, and potentially lower-cost solar devices compared to traditional materials like silicon. Since their introduction in 2009 with efficiencies around 3.8%, single-junction perovskite solar cells have rapidly advanced to over 26% by 2024, as tracked by the *NLR Best Research-Cell Efficiency Chart* (Figure 1.8).

Other key innovations in Si solar cell technology include advancements in cell design, material science, and manufacturing processes. PERC technology, which improves efficiency by reducing electron recombination, is a major advancement, along with TOPCon (Tunnel Oxide Passivated Contact) cells that offer further efficiency gains. Bifacial solar panels, which capture sunlight from both sides, and the development of tandem solar cells (combining Si with materials like perovskite) to broaden the spectrum of light absorbed, are also significant. Furthermore, innovations in light trapping, such as nanostructures, enhanced light absorption, while BIPV and solar windows demonstrate novel applications for Si cells. In addition, tandem solar cells can help to capture more of the sunlight. Tandem combines silicon with other materials, like perovskites, to capture a wider range of the solar spectrum. The perovskite layer absorbs shorter wavelengths, while the silicon absorbs longer wavelengths, leading to higher overall efficiency. In addition, nanostructures and other techniques are used to improve light absorption within the silicon, increasing the amount of sunlight converted to electricity.

6.8 Chapter Summary

1) A solar cell operates as a pn junction device designed to absorb photons, generate electron–hole pairs, separate them before recombining, and finally collect the carriers. The design emphasizes a *p*-type absorber and n-type emitter configuration for maximum performance, since electron mobility is

higher than hole mobility, allowing for higher carrier collection efficiency and reduced recombination losses.

2) The absorber layer is the heart of the solar cell, as it is the location where most of the photons are absorbed, and most of the electron–hole pairs are created. Key properties that determine performance include the bandgap, absorption coefficient, and thickness. The tradeoff between open-circuit voltage and short-circuit current leads to an optimum bandgap of around 1.4 eV for maximum efficiency.
3) The thickness of the absorber layer depends strongly on the absorption coefficient of the material. Direct-bandgap materials such as GaAs and CdTe exhibit short absorption lengths and require only a few micrometers of thickness to absorb most of the solar spectrum, while indirect-bandgap silicon requires much thicker layers, typically on the order of 100–200 micrometers, to achieve similar absorption.
4) The emitter forms the *pn* junction and provides the built-in electric field needed to separate charge carriers. Its thickness and doping concentration are critical design parameters. The emitter must be thin enough to minimize recombination and parasitic absorption, yet thick and heavily doped enough to maintain high conductivity and a strong electric field for efficient carrier separation.
5) Reflection at the front surface of a solar cell represents a major optical loss, with bare silicon reflecting a significant fraction of incident light due to the refractive-index mismatch between air and silicon. ARCs reduce this loss by causing destructive interference between reflected waves, thereby increasing photon transmission into the absorber layer and improving overall efficiency.
6) Metal finger contacts are required to collect charge carriers and reduce series resistance, but they also block incoming light and cause shading losses. An optimized design uses many thin, closely spaced fingers to balance resistance and shading, while TCOs can further improve carrier collection without significantly reducing light absorption.
7) Modern improvements in solar cell performance rely on novel materials and advanced designs. High-absorption materials such as perovskites enable ultra-thin absorber layers, while innovations including PERC, TOPCon, bifacial, and tandem solar cells improve light absorption and reduce recombination losses, leading to higher conversion efficiencies.

End-of-Chapter Problems

P6.1 From Figure 6.3, the absorption length of silicon at 600 nm is approximately 0.9 μm.

a) What fraction of incident 600 nm light is absorbed after one absorption length?
b) What fraction is absorbed after a thickness of 3 μm?

P6.2 Using Figure 6.3, the absorption lengths at 800 nm are approximately:
- Silicon: $L_{Si} = 10\ \mu m$
- GaAs: $L_{GaAs} = 0.7\ \mu m$

Calculate the thickness required to absorb 90% of the incident light for each material.

P6.3 A silicon solar cell has a 50 nm emitter. Using Figure 6.3, estimate the percentage of incident light absorbed in the emitter at:
a) 300 nm
b) 500 nm
Assume: $L_{300} = 0.08\ \mu m$ and $L_{500} = 0.6\ \mu m$

P6.4 You are designing a solar cell with a maximum absorber thickness of 2 μm. Using Figure 6.3, determine which of the following materials can absorb at least 90% of visible light (400–700 nm) within this thickness: Si, GaAs, CdTe, and GaP. Explain your answer qualitatively.

P6.5 Explain why making the emitter extremely thin (e.g. <5 nm) is also undesirable, even though it minimizes parasitic absorption.

P6.6 Explain why materials with high absorption coefficients enable lightweight and flexible solar cells, while silicon generally does not.

P6.7 Explain why commercial solar cells typically use many thin fingers instead of a few wide fingers, even though wide fingers reduce resistance.

P6.8 The ideal refractive index for a single-layer ARC on silicon is:

$$n_{ideal} = \sqrt{n_{air} n_{Si}}$$

a) Calculate the ideal refractive index, n.
b) Compare this value to Si_3N_4 and TiO_2.
c) Which is closer to ideal, and why does this matter?

P6.9 A silicon solar cell without an ARC reflects 31% of incident light at normal incidence. Assume all transmitted photons generate current. If an ARC reduces reflectance to 5%, what is the percentage increase in available photons?

P6.10 Optical losses in a poor-quality silicon solar cell

A silicon solar cell has the following characteristics:

- No ARC. Assume normal-incidence reflectance at the air–Si surface is $R = 0.31$ for given wavelength.
- A 20 nm emitter with very poor carrier lifetime and a poor emitter–absorber interface, so any light absorbed in the emitter is treated as lost (parasitic).
- An absorber (base) thickness of 2 μm.
- Consider monochromatic light at $\lambda = 500$ nm. From Figure 6.3, the absorption length of silicon at 500 nm is $L = 0.6$ μm.
- Assume single-pass absorption and neglect all other optical effects.
  a) Of the light that enters the cell, what fraction is absorbed in the emitter and what fraction passes into the absorber?
  b) Of the light that reaches the absorber, what fraction is absorbed within the 2 μm absorber?

Reference

1 refractiveindex.info, "Refractive index database: optical properties of inorganic materials," [Online]. Available: https://refractiveindex.info/?shelf=main&book=MoS2&page=Islam-o. [Accessed 2026].

7

Efficiency Enhancement Techniques

7.1 About the Chapter

In this chapter, we explore how solar cells can be manufactured to be more efficient. While solar cells are an elegant solution to convert photons to electrons, there are limitations based on the materials available. As discussed before, the bandgap of the semiconductor is a fundamental limitation. The Shockley–Queisser limit, calculated in 1961, sets a maximum theoretical efficiency for a single-junction solar cell at about 33%. To improve efficiency, light management is used where coatings, textures, and nanostructures help capture more sunlight. Next are tandem and multi-junction cells, which layer different materials to harvest a wider range of the solar spectrum. Finally, we introduce emerging PV technologies, such as perovskites, organics, two-dimensional (2D) materials, and quantum dots (QDs), that offer exciting possibilities for lighter, cheaper, and more versatile solar technologies. Together, these approaches provide a glimpse of the innovations that could shape the next generation of solar cells.

Key Topics

- Efficiency limits and the Shockley–Queisser limit
- Light management and surface texturing
- Tandem and multi-junction solar cells
- Emerging photovoltaic technologies
- Concentrated photovoltaic (CPV) systems

7.2 Light Management

The more photons that reach the absorber layer, the more electron–hole pairs can be generated and extracted. Traditionally, anti-reflection coatings are used to minimize the number of reflected photons. In addition to coatings, other methods can be employed to enhance photon absorption within the device. To understand these approaches, it is important to introduce the concept of optical path length (OPL). OPL quantifies the effective distance that light travels inside a medium by accounting for both the geometric thickness of the material and its refractive index. Mathematically, it is defined as

$$L_{opt} = n_s d \tag{7.1}$$

where n_s is the refractive index and d is the physical thickness.

This expression shows that a material with a higher refractive index effectively slows down light, increasing the optical length compared to the actual geometric path. As discussed in Chapter 2, absorption in solar cells is governed by the Beer–Lambert law: $\mathbb{I}(x) = \mathbb{I}_0 e^{-\alpha x}$, where α is the absorption coefficient of the material (from Eq. (2.10)). A longer OPL enhances absorption because photons have a higher probability of being absorbed before escaping the device. Advanced light-trapping designs, therefore, aim to increase the *effective* OPL far beyond the physical thickness of the absorber. Theoretical models, such as the *Yablonovitch limit* [1], show that the maximum path length enhancement in a randomizing medium is

$$L_{eff} \approx 4n_s^2 d \tag{7.2}$$

which represents the ultimate light-trapping limit.

Surface texturing is one of the most effective ways to increase the OPL in solar cells without increasing the physical thickness of the absorber layer. A flat surface allows most photons to propagate nearly straight through the material, giving them essentially a single pass through and only one chance to be absorbed over a distance equal to the thickness. By introducing micro- or nanoscale textures such as pyramids, cones, random surface roughness, or nanoparticles, incident light is refracted and scattered into oblique angles as they enter the cell. As a result, photon average path length increases because the rays travel diagonally through the absorber, effectively making the optical distance longer. Furthermore, some of the reflected light gets reabsorbed, allowing those photons to be absorbed, as illustrated in Figure 7.1.

The textured surfaces promote multiple internal reflections, especially when combined with a reflective back contact. Instead of escaping after one pass, photons can bounce several times inside the absorber, each time increasing the probability of absorption. In fact, under ideal randomizing conditions, the maximum enhancement of the OPL is given by the Yablonovitch limit, which shows that the effective path can be up to $4{n_s}^2$ larger than the physical thickness. This is particularly important for thin-film solar cells, where the absorber layer may be only a few hundred nanometers thick, which is too thin to absorb all incident sunlight in a single pass. Through texturing, these thin layers can perform nearly as well as much thicker wafers while using less material.

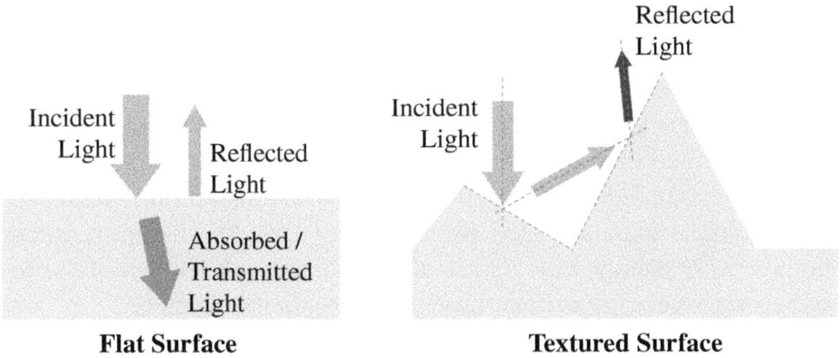

Figure 7.1 Illustration comparing light reflection at (a) flat surface and (b) textured surface, showing how reflected light gets reabsorbed again in textured surface, leading to overall lower reflectance.

Example 7.1 ***Effect of texturing***
A thin-film Si solar cell has an absorber layer thickness of 2 μm with refractive index $n_s = 3.5$.

a) Calculate the optical length for flat surface.
b) If the texturing is done to reach the Yablonovitch limit, calculate the new OPL.
c) If the absorption coefficient is $\alpha = 10^4$ cm^{-1}, find the absorption fraction A for the flat vs. textured surface.
d) Compare the flat to textured surface.

Solution

a) OPL is
   a) $L_{opt} = n_s d = 3.5 \times 2\ \mu\text{m} = 7\ \mu\text{m} = 7 \times 10^{-4}$ cm
b) OPL for textured case:
   a) $L_{opt} \approx 4{n_s}^2 d = 4 \times 3.5^2 \times 2\ \mu\text{m} = 98\ \mu\text{m} = 9.8 \times 10^{-3}$ cm
c) Find the absorption fraction A using Beer–Lambert law:

$$A(L) = 1 - e^{-\alpha L}$$

Flat case (no texture) implies a single pass, where $L = d = 2$ μm

$$A(L) = 1 - e^{-10^4 \times 2 \times 10^{-4}}$$

$$A(L) = 1 - e^{-2} = 0.864 \text{ or } 86.4\%$$

Textured case, where $L = 98$ μm, from the Yablonovitch, we have:

$$A(L) = 1 - e^{-10^4 \times 9.8 \times 10^{-3}}$$

$$A(L) = 1 - e^{-98} \approx 1 \text{ or } 100\%$$

By texturing, the absorption increases from 86.4 to ~100%.

7.3 Tandem and Multi-Junction Cells

Tandem solar cells, also called multi-junction solar cells, are designed to overcome the fundamental efficiency limit of single-junction devices. In a single-junction cell, the maximum theoretical efficiency is given by the Shockley–Queisser limit, which is approximately 33% for a device with an optimum bandgap (i.e. $E_g \sim 1.34$ eV) under standard solar illumination ($AM\,1.5$). This limit arises because photons with energy $h\nu < E_g$ are not absorbed, while photons with $h\nu \geq E_g$ lose their excess energy as heat through carrier thermalization.

7.3.1 Concept and Structure

A tandem solar cell stacks two or more subcells with different bandgaps (E_{g1}, E_{g2}, …) on top of each other. The top subcell absorbs high-energy photons, while lower-energy photons pass through it and are absorbed by the lower-bandgap subcells. This spectral splitting allows a larger fraction of the solar spectrum to be converted into electrical energy more efficiently.

For a two-junction tandem cell, the top subcell with bandgap E_{g1} and the bottom subcell with bandgap E_{g2} each generates current densities J_1 and J_2, and open-circuit voltages V_{oc1} and V_{oc2}, respectively. When the subcells are connected in series, which is the most common configuration, the total current density, J_t, is limited by the smaller of the two

$$J_t = \min(J_1, J_2) \tag{7.3}$$

The total open-circuit voltage of the tandem cell is the sum of the subcell voltages,

$$V_{oc,tandem} = V_{oc1} + V_{oc2} \tag{7.4}$$

The current density generated by each subcell depends on its spectral response $S_i(E)$ and the incident solar photon flux as a function of energy, $\Phi(E)$, where the photon energy is $E = h\nu$.

$$J_i = q \int_{E_{gi}}^{\infty} \Phi(E) S_i(E) dE \tag{7.5}$$

where q is the electron charge. Current matching is a key design challenge in series-connected tandem solar cells since any excess current generated by one subcell cannot be utilized if another subcell limits the total current. In a series-connected tandem solar cell, a subcell producing a lower current limits the operating current of the entire device. The higher-current subcells are forced to operate away from their maximum power point, effectively dissipating excess photogenerated energy internally rather than converting it into useful electrical output. This current mismatch leads to a reduction in the overall conversion efficiency and represents a fundamental design constraint in tandem and multi-junction solar cells.

7.3.2 Theoretical Efficiency

The efficiency, η, of a tandem cell can be calculated as:

$$\eta = \frac{P_{out}}{P_{in}} = \frac{J_t(V_{oc1} + V_{oc2})FF}{P_{in}} \tag{7.6}$$

where FF is the fill factor, and P_{in} is the incident solar power density. The efficiency limit increases with the number of junctions since each can operate near its optimal bandgap.

Figure 7.2 illustrates light absorption in a multi-junction solar cell stack (a), where high-energy photons (short-wavelength light) are absorbed in high-bandgap top subcell, while lower-energy photons are transmitted to and absorbed by the underlying subcells with smaller bandgaps.

Germanium is often used as the bottom subcell in commercial multi-junction solar cells because it has a small bandgap and can absorb low-energy infrared light. Above the germanium layer, subcells made from different $III-V$ semiconductor materials are typically placed. These materials are chosen because their bandgaps can be adjusted and their crystal structure closely matches that of germanium,

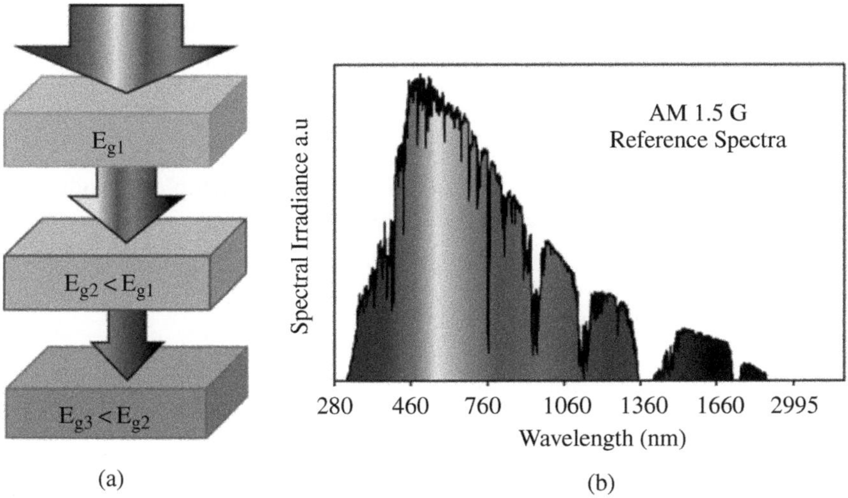

Figure 7.2 (a) Illustration of light absorption in multi-junction solar cell stack and (b) corresponding wavelength distribution in the standard AM 1.5G solar spectrum [2] (Reproduced with permission from Elsevier © 2023, [3]).

which helps reduce defects and thus improves conversion efficiency. However, germanium substrates are expensive and contribute significantly to the overall cost of multi-junction solar cells. For this reason, current research focuses on reducing the cost of multi-junction solar cells, exploring different materials that yield lower costs. One important approach is to replace germanium with cheaper silicon substrates while maintaining good device performance.

Different models in the literature are used to estimate the theoretical upper efficiency limit of multi-junction solar cells connected in series [4–12]. A commonly used approach is the detailed-balance method, which determines the thermodynamic efficiency limit of a given solar cell design [7, 8]. In this method, each subcell is assumed to absorb all photons with energies greater than its bandgap, while photons with energies below the bandgap are transmitted to the underlying layers.

It has been shown in [5] that optimal performance of multi-junction stack can be achieved when the rear surface of each subcell acts as a perfect reflector. This reflector returns the light emitted by a subcell back into the same layer, while still allowing lower-energy photons to pass through to the subcell below. Under these ideal assumptions, detailed-balance calculations predict that the maximum theoretical efficiency of an ideal two-terminal (2*T*) dual-junction (2*J*) solar cell is approximately $42-45\,\%$, for a design with a top-cell bandgap $E_g \sim 1.6-1.64$ eV and a bottom-cell bandgap $E_g \sim 0.94-0.96$ eV [4, 9, 10]. The top cell absorbs visible light efficiently, while the bottom silicon cell captures near-infrared photons.

The exact optimum bandgap values and maximum efficiencies reported in literature vary slightly, depending on the model and assumptions used in calculations. Using similar approaches, the optimum bandgaps for a three-junction solar cell have been estimated to be approximately 1.90 eV/1.36 eV/0.94 eV for the top, middle, and bottom subcells, respectively [4, 9].

Figure 7.3 shows upper efficiency limit of two-terminal dual-junction solar cell under *AM* 1.5G incident spectrum, as a function of bottom- and top-cell bandgaps. For conventionally designed 2*J* tandem cell, maximum efficiency limit is ~46% for 1.64 eV/0.96 eV top- and bottom-cell bandgap under *AM* 1.5G.

Example 7.2 ***Tandem solar cell***

If you have two junction tandem cell with the following details:

- Top subcell: short-circuit current density = 16 mA/cm^2 with open circuit of voltage = 1.62 V
- Bottom subcell: short-circuit current density = 22 mA/cm^2 with open circuit of voltage = 0.98 V
- Cells are connected in series and with same area, *A*

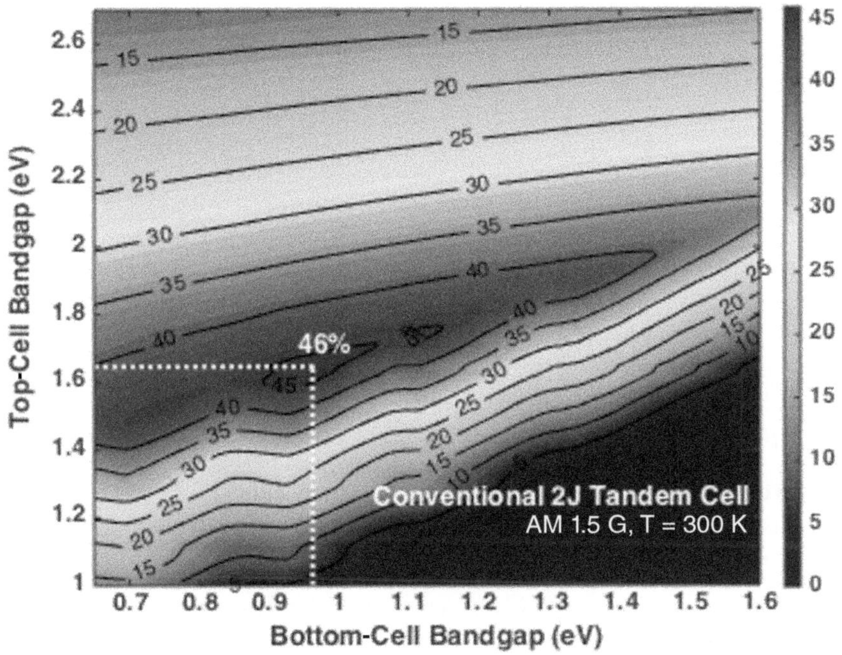

Figure 7.3 Efficiency of a 2J traditionally designed tandem cell as a function of top- and bottom-cell bandgap under AM 1.5G incident spectra ([2], reproduced with permissions from Elsevier © 2023, [13]).

- Fill-Factor is 80%
- Incident Power Density = 1000 W/m^2
- Area = 4 cm^2

a) Determine the current density of the tandem
b) Find the tandem open-circuit voltage
c) Find the output power density
d) Calculate the efficiency
e) Which subcell limits the tandem cell current
f) You add an anti-reflection coating, make the top cell current 20 mA/cm^2, repeat a–e

Solution

a) $J_t = \min(J_1, J_2) = \min(16 \text{ mA/cm}^2, 22 \text{ mA/cm}^2) = 16 \text{ mA/cm}^2$
b) $V_{oc,tandem} = V_{oc1} + V_{oc2} = 1.62 \text{ V} + 0.98 \text{ V} = 2.6 \text{ V}$

c) $P_{out} = FF \times J_{sc} \times V_{oc}$

$$P_{out} = 0.8 \times 16\ \text{mA/cm}^2 \times 2.6\ \text{V} = 33.28\ \text{mW/cm}^2$$

$$= 33.28 \frac{W}{\text{cm}^2} \times 4\ \text{cm}^2$$

$$= 133.2\ \text{mW}$$

d) $\eta = \frac{P_{out}}{P_{in}}$

$$P_{in} = 100\ \text{mW/cm}^2\ \ (\text{for } AM\ 1.5\ G)$$

$$= 100\ \text{mW/cm}^2 \times 4\ \text{cm}^2$$

$$P_{in} = 400\ \text{mW}$$

$$\eta = \frac{133.2}{400} = 0.33 = 33.3\%$$

e) The top cell limits the current due to the series connection

f) You add an anti-reflection coating, make the top cell current 20 mA/cm² repeat *a–e*, assume that V_{oc} did not change.

$$J_t = \min(J_1, J_2) = \min(20\ \text{mA/cm}^2, 22\ \text{mA/cm}^2) = 20\ \text{mA/cm}^2$$

$$V_{oc,tandem} = V_{oc1} + V_{oc2} = 1.62\ \text{V} + 0.98\ \text{V} = 2.6\ \text{V}$$

$$P_{out} = FF \times J_{sc} \times V_{oc}$$

$$P_{out} = 0.8 \times 20\ \text{mA/cm}^2 \times 2.6\ \text{V} = 41.6\ \ \text{W/cm}^2$$

Output power density

$$P_{out} = 41.6 \frac{W}{\text{cm}^2} \times 4\ \text{cm}^2$$

$$P_{out} = 166.4\ \text{mW}$$

Output power

$$\eta = \frac{P_{out}}{P_{in}}$$

$$P_{in} = 100\ \text{mW/cm}^2\ (AM\ 1.5\ G) = 100\ \text{mW/cm}^2 \times 4\ \text{cm}^2 = 400\ \text{mW}$$

$$\eta = \frac{166.4}{400} = 0.33 = 41.6\%$$

7.4 Emerging Photovoltaic Technologies—Beyond Silicon

While silicon remains the dominant material for commercial solar cells, researchers continue exploring new materials that can capture sunlight more efficiently, use less energy to manufacture, or enable flexible and lightweight designs. These emerging PV technologies are still developing but show exciting potential, especially when combined with silicon in tandem structures.

7.4.1 Perovskite Solar Cells

Metal halide perovskites are thin-film materials that absorb light very efficiently and have tunable bandgaps (typically 1.5–1.8 eV). As a result, perovskites are ideal top cells in tandem with silicon ($E_g \approx 1.1$ eV). The perovskite layer captures visible light, while silicon absorbs the near-infrared part of the spectrum. Recent lab devices have reached efficiencies above 30 %, showing that perovskite–Si tandems can surpass the single-junction silicon limit (Figure 1.8). While perovskites show incredible promise for solar with high absorption, low-cost processing, and spectrum tunability, the major challenges are long-term stability and moisture sensitivity, which could derail their deployment.

7.4.2 Organic Solar Cells

Organic solar cells (OSCs) use organic semiconductors made from carbon-based molecules or polymers that can be printed on flexible films. One of their key features is that these materials can be processed from liquid solutions, which makes it possible to print solar cells onto thin, lightweight, and flexible films. The main advantage of OSCs is that, due to their lightweight structure, they have potential for roll-to-roll fabrication, utilizing fabrication methods like printing newspapers, significantly lowering production costs.

In most OSCs, light absorption takes place in a thin active layer composed of two different organic materials: a donor and an acceptor. When light is absorbed, it creates tightly bound electron–hole pairs called excitons. Unlike in inorganic semiconductors, these excitons do not immediately separate into free charges. Instead, they must first reach the interface between the donor and acceptor materials, where charge separation can occur. This requirement places strong constraints on the device structure and limits how thick the active layer can be.

Modern OSCs reach over 20% efficiency in the lab. However, they typically degrade faster under sunlight and oxygen exposure. As a result, their operational lifetime is typically shorter than that of silicon-based solar cells. In addition, charge carriers move more slowly in organic semiconductors, which can lead to higher recombination losses and reduced performance. Current research in organic PVs, therefore, focuses not only on improving efficiency, but also on enhancing material stability and developing effective encapsulation techniques to extend device lifetime.

7.4.3 Two-Dimensional Materials

2D materials such as molybdenum disulfide (MoS_2) and tungsten disulfide (WS_2) are only a few atoms thick, yet they can interact very strongly with light. Even though these materials are extremely thin, they exhibit relatively high optical absorption at certain wavelengths due to their electronic structure. This unusual combination of atomic-scale thickness and strong light–matter interaction has made 2D materials an active area of research in PVs and optoelectronics.

One of the key advantages of 2D materials is their ability to form atomically sharp interfaces with other semiconductors, such as silicon or perovskite absorber layers. Since their surfaces are naturally free of dangling bonds, they are valuable as passivation layers or charge-selective contacts that improve voltage and reduce recombination.

While 2D materials, such as MoS_2 and WS_2, can act as light absorbers, they are often employed as charge-selective contacts or passivation layers to enhance efficiency. In these roles, their specific energy band alignments allow them to selectively transport holes while blocking electrons (or vice versa), which significantly reduces recombination losses at the device interfaces. Though not yet practical as main absorbers, they play a growing role in interface engineering.

7.4.4 Quantum-Dot Solar Cells

QDs are nanocrystals whose bandgap depends on their size—smaller dots have higher bandgaps. They can be tuned to absorb parts of the spectrum that silicon misses (especially near-infrared). QDs are also being explored as infrared-harvesting layers in tandem cells. While QDs provide tunable absorption, solution processing, and compatibility with flexible substrates, they still suffer from material stability and surface defect control. Table 7.1 highlights the emerging PV technologies beyond silicon.

Table 7.1 Emerging photovoltaic technologies—Beyond silicon.

Material type	Typical bandgap (eV)	Key advantage	Main challenge	Status
Perovskite	1.5–1.8	High efficiency, tunable E_g	Stability	Commercial pilot
Organic	1.6–2.0	Flexible, low-cost	Degradation	Research-scale
2D materials	1.2–2.0	Ultrathin, good interfaces	Fabrication	Early research
Quantum dots	0.8–2.0	Tunable, IR absorption	Surface traps	Research-scale

Example 7.3 ***Bandgap and absorption edge***

a) A perovskite top cell has a bandgap of 1.7 eV. Calculate the corresponding absorption edge wavelength (in nm).
b) Compare this with silicon ($E_g = 1.1$ eV).
c) Explain why combining these two materials in a tandem structure improves overall efficiency.

Solution

The photon energy ($h\nu$) and wavelength are related by (Eq. (2.1))

$$h\nu = \frac{hc}{\lambda}$$

Using $h = 6.626 \times 10^{-34}$ Js and $c = 3 \times 10^8$ m/s.

$$hc = 6.626 \times 10^{-34} \cdot 3 \times 10^8 = 1.986 \times 10^{25} \text{J} \cdot \text{m}$$

Converting: $1 \text{ eV} = 1.602 \times 10^{-19}$ J, simplify:

$$hc = \frac{1.968 \times 10^{25}}{1.602 \times 10^{-19}} = 1.24 \times 10^{-6} \text{ eV/m}$$

convert to nm

$$hc = 1.24 \times 10^{-6} \text{ eV/m} = 1240 \text{ eV/nm}$$

so

$$\lambda \text{ (nm)} \approx 1240/h\nu(\text{eV}) \tag{7.7}$$

a) For perovskite with $E_g = 1.7$ eV:

$$\lambda = 1240/1.7 = 730 \text{ nm}$$

b) For silicon with $E_g = 1.1$ eV:

$$\lambda = 1240/1.1 = 1127 \text{ nm}$$

c) Therefore, the perovskite absorbs visible light up to about 730 nm, while silicon captures the near-infrared up to about 1127 nm. Stacking these materials in tandem allows each to absorb a different part of the solar spectrum, reducing energy losses and increasing overall conversion efficiency.

7.4.5 Concentrated Solar Applications

Concentrated PV (CPV) power today utilizes low-cost concentrating optics that reduce solar module area required to generate electricity. Under solar concentration, current generated by the cell scales linearly with solar concentration factor, X, of the input spectrum. Due to an increase in optical current, open-circuit voltage increases by the term $ln(X)$; hence, overall efficiency increases. Efficiency does not increase indefinitely, as there is a physical limit to open-circuit voltage, which is the built-in voltage of the cell. Also, with increasing current, fill factor starts to decrease as series resistance effects begin to dominate the IV characteristics. However, benefit of CPVs is not in the efficiency increase, although that is an added advantage, but in the increased power density ($J \cdot V/Area$) generated by the cell. The added cost of concentrating optics and tracking systems should be compensated by the decrease in module area required to generate an equal amount of power. This added advantage made high-efficiency but high-cost multi-junction solar cells somewhat economically feasible. High-concentration systems (between $300X$ and $1000X$) usually utilize $III - V$ materials, while low-concentration (below $100X$) CPV plants are based on silicon solar cells [14].

Today's record efficiencies are held by Sharp and Fraunhofer Institute at 44.4% and 46 %, for triple- and four-junction solar cells at 302 and 508 suns, respectively [15, 16]. However, global CPV capacity is still relatively low, and there has been a decline in yearly added installation of CPV. One of the reasons could be the loss of CPV manufacturing capacity due to closure of the companies or their shift to PV technology. However, CPV market is estimated to have a bright future, with estimates of its cost equaling PV system by year 2030 [14].

CPV systems use optical components such as lenses or mirrors to focus sunlight onto a small-area solar cell, reducing the amount of active semiconductor material required to generate electricity [14]. The degree of concentration is described by the concentration factor X, which indicates how many times the incident solar intensity exceeds standard one-sun illumination.

Under concentrated illumination, the photocurrent generated by a solar cell increases approximately linearly with the concentration factor. At the same time, the open-circuit voltage increases logarithmically with X, leading to an overall improvement in conversion efficiency [7, 8]. This efficiency gain, however, is limited by physical constraints. At high-concentration levels, the open-circuit voltage approaches the built-in potential of the junction, while increased current density enhances resistive losses in the device. These losses reduce the fill factor and limit further efficiency improvements [4, 6].

The primary benefit of CPV systems is therefore not only higher efficiency, but significantly higher power density, defined as the electrical power generated per unit cell area. By producing more power from a smaller cell area, CPV enables the use of high-efficiency but high-cost multi-junction solar cells. For CPV to be economically viable, the additional cost of concentrator optics and solar tracking systems must be offset by the reduced semiconductor area needed to achieve a given power output [14].

High-concentration CPV systems, typically operating between 300× and 1000× sunlight, almost exclusively employ *III–V* multi-junction solar cells due to their superior performance under intense illumination [17]. Lower concentration systems, generally below 100 ×, may utilize silicon solar cells, although the efficiency benefits in this regime are more modest [14].

Record efficiencies reported in the literature demonstrate the strong performance potential of CPV technology. Multi-junction *III–V* solar cells under concentrated illumination have achieved conversion efficiencies exceeding 44% for triple-junction devices and above 46% for four-junction devices under several hundred suns [16].

According to the NLR efficiency charts (Figure 1.8) and the most recent solar cell efficiency tables, multi-junction *III–V* solar cells under concentrated illumination continue to hold the highest confirmed PV efficiencies reported to date [18, 19]. At the same time, comprehensive reviews of CPV technology emphasize that despite these record efficiencies, widespread deployment remains limited due to system complexity, tracking requirements, and strong cost reductions in conventional silicon PVs [20]. As a result, CPV is generally viewed as a high-performance niche technology rather than a mainstream solution for large-scale power generation.

7.5 Chapter Summary

1) The bandgap of the semiconductor is a fundamental limitation to solar cell efficiency, and the Shockley–Queisser limit sets a maximum theoretical efficiency for a single-junction solar cell at about 33% under standard solar illumination.
2) Light management techniques are used to improve efficiency by capturing more sunlight, where coatings, textures, and nanostructures help increase the OPL and enhance photon absorption within the absorber layer.
3) Surface texturing increases the effective OPL by refracting and scattering incident light into oblique angles, promoting multiple internal reflections and allowing thin absorber layers to perform nearly as well as much thicker wafers.
4) Tandem and multi-junction solar cells overcome the efficiency limit of single-junction devices by stacking subcells with different bandgaps, where high-energy photons are absorbed in the top subcell and lower-energy photons are transmitted to and absorbed by lower-bandgap subcells.
5) In series-connected tandem solar cells, the total current density is limited by the subcell generating the smallest current, making current matching a key design challenge that directly affects overall conversion efficiency.

6) Theoretical efficiency limits of multi-junction solar cells increase with the number of junctions, and detailed-balance calculations predict efficiencies exceeding 40% for optimally designed dual- and multi-junction solar cells under ideal conditions.
7) Emerging PV technologies such as perovskite, organic, 2D, and QD solar cells offer new opportunities for high absorption, tunable bandgaps, lightweight structures, and novel device architectures beyond silicon.
8) CPV systems use optical components to focus sunlight onto small-area solar cells, increasing current density and power density and enabling the use of high-efficiency multi-junction solar cells despite their higher cost.

End-of-Chapter Problems

Conceptual Questions

P7.1 Explain why increasing OPL can improve absorption without increasing absorber thickness.

P7.2 Why does surface texturing reduce reflectance compared to a flat surface?

P7.3 State the physical meaning of the Yablonovitch limit.

P7.4 Why is current matching critical in series-connected tandem solar cells?

P7.5 Why are perovskites attractive as top cells in tandem structures with silicon?

True or False

P7.6 Increasing refractive index increases optical path length for a fixed thickness.
True/False

P7.7 The Shockley–Queisser limit applies equally to multi-junction solar cells.
True/False

P7.8 In a series-connected tandem cell, the total voltage is the sum of subcell voltages.
True/False

P7.9 Surface texturing is mainly beneficial for very thick absorber layers.
True/False

P7.10 In CPV systems, current increases approximately linearly with concentration factor.
True/False

Multiple Choice Questions

P7.11 The Yablonovitch light-trapping limit scales as
A $n_s d$
B $2n_s d$
C $4n_s^2 d$
D $n_s^2 d$

P7.12 In a two-junction series-connected tandem cell, the operating current density equals
A $J_1 + J_2$
B $(J_1 + J_2)/2$
C $\max(J_1, J_2)$
D $\min(J_1, J_2)$

P7.13 Which loss mechanism is primarily reduced by tandem solar cells compared to single-junction cells?
A Series resistance loss
B Thermalization loss
C Optical reflection loss
D Contact recombination loss

P7.14 Which material property enables QDs to have tunable bandgaps?
A Crystal orientation
B Film thickness
C Nanocrystal size
D Doping concentration

P7.15 The main benefit of CPV systems is
A Lower operating temperature
B Higher fill factor
C Higher power density per unit cell area
D Elimination of tracking systems

Numerical Problems

P7.16 Optical Path Length
A thin-film absorber has thickness $d = 1.5\,\mu\text{m}$ and refractive index $n_s = 3.0$.
a) Calculate the optical path length for a flat surface.
b) Calculate the effective optical path length at the Yablonovitch limit.

P7.17 Absorption Comparison
For the absorber in P7.16, assume $\alpha = 8 \times 10^3\,\text{cm}^{-1}$. Calculate the absorption fraction for
a) flat surface
b) textured surface at the Yablonovitch limit.

P7.18 Tandem Cell Current and Voltage
A tandem solar cell has: Top cell: $J_1 = 18\,\text{mA/cm}^2$, $V_{oc1} = 1.6\,\text{V}$ and Bottom cell: $J_2 = 24\,\text{mA/cm}^2$, $V_{oc2} = 1.0\,\text{V}$.
a) Find the tandem current density.
b) Find the tandem open-circuit voltage.

P7.19 Tandem Efficiency
Consider tandem cell from problem P 7.18. Assume $FF = 0.78$ and $P_{in} = 100\,\text{mW/cm}^2$. Calculate the efficiency of the tandem cell.

P7.20 Bandgap and Absorption Edge
Calculate the absorption edge wavelength for a semiconductor with $E_g = 1.55\,\text{eV}$.

References

1 Yablonovitch, E. (1982). Statistical ray optics. *Journal of the Optical Society of America* 72: 899–907.
2 S. Abdul Hadi, *III–V on Si Multi-Junction Step-Cell: PhD Dissertation*, Abu Dhabi: Masdar Institute of Science and Technology (currently Khalifa University), 2016.
3 Nayfeh, A. and Abdul Hadi, S. (2023). Chapter 3: Basics of solar cells. In: Solar Cell Engineering: Silicon-Germanium Alloys for Photovoltaic Applications, 17–35. Elsevier.

4 Brown, A.S. and Green, M.A. (2002). Detailed balance limit for the series constrained two terminal tandem solar cell. *Physica E: Low-dimensional Systems and Nanostructures* 14 (1–2): 96–100.

5 Marti, A. and Arauijo, G.L. (1996). Limiting efficiencies for photovoltaic energy conversion in multigap systems. *Solar Energy Materials and Solar Cells* 43: 203–222.

6 Kurtz, S., Faine, P., and Olson, J. (1990). Modeling of two-junction, series-connected tandem solar cells using top-cell thickness as an adjustable parameter. *Journal of Applied Physics* 68: 1890.

7 Shockely, W. and Queisser, H.J. (1961). Detailed balance limit of efficiency of p-n junction solar cells. *Journal of Applied Physics* 32: 510.

8 Henry, C. (1980). Limiting efficiencies of ideal single and multiple energy gap terrestrial solar cells. *Journal of Applied Physics* 51: 4494.

9 Feser, C., Lacombe, J., Maydel, K. et al. (2012). A simulation study towards a new concept for realization of thin film triple junction solar cells based on group IV elements. *Progress in Photovoltaics: Research Application* 20: 74–81.

10 Connolly, J.P., Mencaraglia, D., Renard, C. et al. (2014). Designing III–V multijunction solar cells on silicon. *Progress in Photovoltaics: Research Applications* 22: 810–820.

11 Kiess, H. and Rehwald, W. (1995). On the ultimate efficiency of solar cells. *Solar Energy Materials and Solar Cells* 38: 45–55.

12 Vos, A.D. (1980). Detailed balance limit of the efficiency of tandem solar cells. *Journal of Physics D: Applied Physics* 13: 83946.

13 Nayfeh, A. and Abdul Hadi, S. (2023). Chapter 7 – Modeling and simulations of Si_{1-x} Ge_x based solar cells. In: Silicon-Germanium Alloys for Photovoltaic Applications – in Solar Cell Engineering, 103–148. Elsevier.

14 S. Philipps, A. W. Bett, K. Horowitz et al., "Current status of concentrator photovoltaics (CPV) technology," 2016. [Online]. Available: https://www.ise.fraunhofer.de. [Accessed 2016].

15 National Renewable Energy Laboratory (NREL), "Renewable resource data center," [Online]. Available: http://rredc.nrel.gov/solar/spectra/am1.5/. [Accessed 2026].

16 Green, M.A., Emery, K., Hishikava, Y. et al. (2016). Solar cell efficiency tables (version 47). *Progress Photovoltaics: Research Applications* 24: 3–11.

17 Abdul Hadi, S., Fitzgerald, E.A., and Nayfeh, A. (2016). Theoretical efficiency limit for a two-terminal multi-junction "step-cell" using detailed balance method. *Journal of Applied Physics* 119: 073104.

18 National Renewable Energy Laboratory (NREL) (2024). Best Research-Cell Efficiency Chart. National Laboratory of the Rockies (NLR).

19 Green, M.A., Dunlop, E.D., Hohl-Ebinger, J. et al. (2024). Solar cell efficiency tables (latest version). *Progress in Photovoltaics: Research and Applications* 2023–2024: 425–441.

20 S. Kurtz, "Opportunities and challenges for development of a mature concentrating photovoltaic power industry," (p. 5). Golden, CO: National Renewable Energy Laboratory. "It is technical report, published in 2012 by NREL, vol. 2, 2012.

Appendix A

List of Equations

Table A.1 List of key equations in Chapter 2.

$h\nu = \frac{hc}{\lambda}$	(2.1)	$\Phi(\lambda) = S_\lambda(\lambda)\frac{\lambda}{hc}$	(2.2)
$\text{EQE}(\lambda) = \frac{\text{number of collected carriers per second}}{\text{number of incident photons per second}}$	(2.3)	$\text{EQE} \approx (1 - R)IQE$	(2.4)
$J_{ph} = q\int_{\lambda_{\min}}^{\lambda_{\max}} \Phi(\lambda)\, EQE(\lambda)\, d\lambda$	(2.5)	$\tilde{n}(\lambda) = n_s(\lambda) + j\kappa(\lambda)$	(2.6)
$n_s = c_0/c$	(2.7)	$\alpha(\lambda) = \frac{4\pi\kappa(\lambda)}{\lambda}$	(2.8)
$\delta = \frac{1}{\alpha}$	(2.9)	$\mathbb{I}(x) = \mathbb{I}_0 e^{-\alpha x}$	(2.10)
$R(\lambda) + A(\lambda) + \mathscr{T}(\lambda) = 1$	(2.12)	$n_1 \sin\theta_1 = n_2 \sin\theta_2$	(2.13)
$r_{12} = \frac{n_1 - n_2}{n_1 + n_2}$	(2.14)	$R = \|r_{12}\|^2$	(2.15)
$t_{12} = \frac{2n_1}{n_1 + n_2}$	(2.16)	$\mathscr{T} = \frac{n_2}{n_1}\|t_{12}\|^2$	(2.17)
$\mathscr{T} = \frac{n_2}{n_1}\|t\|^2 = \frac{4n_1 n_2}{(n_1 + n_2)^2}$			(2.18)
$M_{\text{prop},i} = \begin{bmatrix} e^{-j\varphi_i} & 0 \\ 0 & e^{j\varphi_i} \end{bmatrix}, \quad \varphi_i = n_i d_i \frac{2\pi}{\lambda}$			(2.20)
$M_{i/(i+1)} = \frac{1}{2n_{(i+1)}} \begin{bmatrix} n_{(i+1)} + n_i & n_{(i+1)} - n_i \\ n_{(i+1)} - n_i & n_{(i+1)} + n_i \end{bmatrix}$			(2.21)
$M = \begin{bmatrix} A & B \\ C & D \end{bmatrix} = \prod_{i=1}^{N} (M_{(i-1)/i}\, M_{\text{prop},i}\, M_{i/(i+1)})$			(2.22)

(Continued)

Table A.1 (Continued)

$R_{0,(N+1)} = \left\lvert \frac{-C}{D} \right\rvert^2$	(2.23)	$2\varphi = \pi \Rightarrow n_1 d = \frac{\lambda_0}{4}$	(2.24)
$n_1 \approx \sqrt{n_0 n_s}$	(2.25)	$\varphi_i = \tilde{n}_i d_i \frac{2\pi}{\lambda}$	(2.26)
$M_{\text{prop},i} = \begin{bmatrix} e^{-j\tilde{n}_i d_i \frac{2\pi}{\lambda}} & 0 \\ 0 & e^{j\tilde{n}_i d_i \frac{2\pi}{\lambda}} \end{bmatrix}$			(2.27)
$M_{i/(i+1)} = \frac{1}{2\tilde{n}_{(i+1)}} \begin{bmatrix} \tilde{n}_{(i+1)} + \tilde{n}_i & \tilde{n}_{(i+1)} - \tilde{n}_i \\ \tilde{n}_{(i+1)} - \tilde{n}_i & \tilde{n}_{(i+1)} + \tilde{n}_i \end{bmatrix}$			(2.28)
$R(\lambda) = \lvert r(\lambda) \rvert^2$	(2.29)	$\mathscr{T}(\lambda) = \frac{n_{out}}{n_{in}} \lvert t(\lambda) \rvert^2$	(2.30)

Table A.2 List of key equations in Chapter 3.

$n = p = \text{EHPs (cm}^{-3})$	(3.1)	$G = \mathcal{R}$	(3.2)
$n = p = n_i$	(3.3)	$np = n_i^2$	(3.4)
$n_i(T) = BT^{3/2} e^{-E_g/2kT}$	(3.5)	$n_0 \cong N_d$	(3.6)
$p_0 \cong N_a$	(3.7)	$f_F(E) = \frac{1}{1 + e^{(E-E_f)/kT}}$	(3.8)
$n(E) = \int_{E_c}^{\infty} f_F(E)\, N(E)\, dE$	(3.9)	$n(E_c) = N_c f_F(E_c)$	(3.10)
$f_F(E_C) = \frac{1}{1 + e^{(E_c-E_f)/kT}} \approx e^{-(E_c-E_f)/kT}$			(3.11)
$n_0 = N_c\, e^{-(E_f-E_c)/kT}$			(3.12)
$f_F(E_v) = \frac{1}{1 + e^{(E_v-E_f)/kT}} \approx e^{-(E_f-E_v)/kT}$			(3.13)
$p = \int_{E_v}^{\infty} f_F(E)\, N(E)\, dE$	(3.14)	$p_0 = N_v\, e^{-(E_f-E_v)/kT}$	(3.15)
$n_0 p_0 = N_c N_v e^{-(E_g/kT)}$	(3.16)	$n_i = \sqrt{N_c N_v e^{-(E_g/kT)}}$	(3.17)
$n_0 p_0 = n_i^2$	(3.18)	$n_i = N_c e^{-(E_c-E_i)/kT}$	(3.19)
$n_0 = n_i e^{(E_f-E_i)/kT}$	(3.20)	$p_i = n_i = N_v e^{-(E_i-E_v)/kT}$	(3.21)
$p_0 = p_i e^{(E_i-E_f)/kT} = n_i e^{(E_i-E_f)/kT}$			(3.22)

Table A.2 (Continued)

$n_0 = N_d - N_a$	(3.26)	$p_0 = N_a - N_d$	(3.27)
$F = q\text{E} = m^* a$	(3.30)	$\mu_n = \dfrac{q\bar{t}}{m_n^*}$	(3.43)
$\mu_p = \dfrac{q\bar{t}}{m_p^*}$	(3.44)	$J_p = qp\mu_p\text{E}$	(3.45)
$J_n = qn\mu_n E$	(3.46)	$J_{total} = J_n + J_p$	(3.47)
$J_{total} = (n\mu_n + p\mu_p)qE = \sigma E$	(3.48)	$\sigma = (n\mu_n + p\mu_p)q$	(3.48)
$R = \rho\dfrac{L}{A}$	(3.49)	$R = \dfrac{L}{(n\mu_n + p\mu_p)qA}$	(3.50)
$\dfrac{1}{\mu_{total}} = \dfrac{1}{\mu_L} + \dfrac{1}{\mu_I} + \ldots$	(3.51)	$\text{E} = -\dfrac{dV}{dX}$	(3.52)
$\Delta n(t) = \Delta n(0)e^{-t/\tau_n}$	(3.53)	$\Delta p(t) = \Delta p(0)e^{-t/\tau_p}$	(3.54)
$\delta n = \tau_n \cdot g_{op}$ and $\delta p = \tau_p \cdot g_{op}$	(3.55)	$n = n_0 + \delta n$ and $p = p_0 + \delta p$	(3.56)
$n = n_i e^{\frac{(F_n - E_i)}{kT}}$	(3.57)	$p = n_i e^{\frac{(E_i - F_p)}{kT}}$	(3.58)
$D_{n,p} \equiv \dfrac{\bar{l}^2}{\bar{\tau}}$	(3.59)	$J_n = qD_n\dfrac{dn(x)}{dx}$	(3.60)
$J_p = -qD_p\dfrac{dp(x)}{dx}$	(3.61)	$\Delta p(x) = \Delta p(0)e^{-x/L_p}$	(3.62)
$L_p = \sqrt{D_p\tau_p}$	(3.63)	$\Delta n(x) = \Delta n(0)e^{-x/L_n}$	(3.64)
$L_n = \sqrt{D_n\tau_n}$	(3.65)	$J_{total} = J_n + J_p$	(3.66)
$J_{n/p} = J_{drift} + J_{diffusion}$	(3.67)	$J_n = q\mu_n n(x)\text{E}_x + qD_n\dfrac{dn(x)}{dx}$	(3.68)
$J_p = q\mu_p p(x)\text{E}_x - qD_p\dfrac{dp(x)}{dx}$	(3.69)	$\dfrac{D}{\mu} = \dfrac{kT}{q}$	(3.70)

Table A.3 List of equations in Chapter 4.

$V_0 = (E_f - E_i)_{n\text{-}side} + (E_i - E_f)_{p\text{-}side}$			(4.1)
$kT \times ln\frac{n_0}{n_i} = E_f - E_i$	(4.2)	$V_0 = \frac{kT}{q} ln\frac{N_d N_a}{n_i^{\ 2}}$	(4.13)
$Q_n^+ = qAx_n N_d$	(4.14)	$Q_p^- = qAx_p N_a$	(4.15)
$x_n N_d = x_p N_a$			(4.16)
$\text{E}(0) = \text{E}_0 = \text{E}_{\max} = -\frac{q}{\varepsilon} N_d x_0 = -\frac{q}{\varepsilon} N_a x_{p0}$			(4.28)
$-V_0 = \frac{1}{2} W \text{E}_0$	(4.32)	$V_0 = \frac{1}{2}\frac{q}{\varepsilon}\frac{N_a N_d}{N_a + N_d} W^2$	(4.36)
$V_0 = -\frac{1}{2} W \frac{q}{\varepsilon} N_d x_{n0} = \frac{1}{2} W \frac{q}{\varepsilon} N_a x_{p0}$			(4.33)
$x_n = W\left(\frac{N_a}{N_a + N_d}\right)$	(4.34)	$x_p = W\left(\frac{N_d}{N_a + N_d}\right)$	(4.35)
$W = \sqrt{\frac{2\varepsilon(V_0 - V_a)}{q}\left(\frac{1}{N_a} + \frac{1}{N_d}\right)}$	(4.37)	$W = \sqrt{\frac{2\varepsilon kT}{q^2} ln\frac{N_a N_d}{n_i^2}\left(\frac{1}{N_a} + \frac{1}{N_d}\right)}$	(4.38)
$V_{RB} = V_0 + \lvert V_a \rvert$	(4.39)	$V_{FB} = V_0 - \lvert V_a \rvert$	(4.40)
$I = I_0\left(e^{qV_a/kT} - 1\right)$	(4.41)	$I_0 = qA\left(\frac{D_p}{L_p} p_n + \frac{D_n}{L_n} n_p\right)$	(4.42)
$I = \left(\frac{D_p}{L_p} p_n + \frac{D_n}{L_n} n_p\right)\left(e^{-qV_a/kT} - 1\right)$			(4.43)
$I = \left(\frac{D_p}{L_p} p_n\right)\left(e^{-qV_a/kT}\right)$	(4.44)	$I = \left(\frac{D_p}{L_p} N_v\right)\left(e^{[qV_a - (E_{fn} - E_{fv})/kT]}\right)$	(4.45)
$I = I_0(e^{qV/\text{nkT}} - 1)$			(4.46)

Table A.4 List of equations in Chapter 5.

$I_p = qAg_{op}(W + L_n + L_p)$	(5.1)
$I_t = I_d + I_p$	(5.2)
$I_t = I_0\left(e^{qV_a/kT} - 1\right) + q \cdot A \cdot g_{op}(W + L_n + L_p)$	(5.3)
$I_t = I_0\left(e^{qV_a/kT} - 1\right) - I_p$	(5.4)
$V_{oc} = \frac{kT}{q} ln\left(\frac{I_p}{I_0} + 1\right)$	(5.5)
$V_{oc} \approx \frac{kT}{q} ln\left(\frac{I_p}{I_0}\right)$	(5.6)
$J_{sc} = I_p/A$	(5.8)
$\left(e^{qV_m/kT}\right)\left(1 + V_m \frac{q}{kT}\right) = \frac{I_0 + I_p}{I_0}$	(5.10)
$FF = \frac{J_m V_m}{J_{sc} V_{oc}}$	(5.11)
$P_{\max} = J_m V_m$	(5.12)
$P_{\max} = FF \cdot J_{sc} \cdot V_{oc}$	(5.13)
$\eta = \frac{P_{\max}}{P_{in}}$	(5.14)
$\eta = \frac{FF \cdot J_{sc} \cdot V_{oc}}{P_{in}}$	(5.15)

Table A.5 List of key equations in Chapter 7.

$L_{opt} = n_s d$	(7.1)
$L_{eff} \approx 4n_s^2 d$	(7.2)
$J_t = \min(J_1, J_2)$	(7.3)
$V_{oc,tandem} = V_{oc1} + V_{oc2}$	(7.4)
$J_i = q \int_{E_{gi}}^{\infty} \phi(E) S_i(E) dE,\ E = h\nu$ (photon energy)	(7.5)
$\eta = \frac{P_{out}}{P_{in}} = \frac{J_t(V_{oc1} + V_{oc2})FF}{P_{in}}$	(7.6)
$\lambda\,(\text{nm}) \approx 1240/h\nu\,(\text{eV})$	(7.7)

Appendix B

Physical Constants, Unit Conversions, and Parameters

Table B.1 Key physical constants.

Constant	Value
Boltzmann's constant	$k = 8.62 \times 10^{-5}$ eV/K
Electronic charge (magnitude)	$q = 1.60 \times 10^{-19}$ C
Electronic rest mass	$m_0 = 9.11 \times 10^{-31}$ kg
Permittivity of free space	$\varepsilon_0 = 8.85 \times 10^{-14}$ F/cm
Planck's constant	$h = 4.14 \times 10^{-15}$ eV · s
Room temperature value of kT	$kT = 0.0259$ eV
Speed of light	$c = 2.998 \times 10^{10}$ cm/s

Table B.2 Unit conversions.

Quantity	Equivalent value
1 Å (angstrom)	10^{-8} cm
1 μm (micron)	10^{-4} cm
1 nm	10 Å = 10^{-7} cm
2.54 cm	1 in
1 eV	1.6×10^{-19} J

Table B.3 Si, GaAs, and Ge properties ($T = 300$ K).

Property	Si	GaAs	Ge
Dielectric constant	11.7	13.1	16.0
Bandgap energy (eV)	1.12	1.42	0.66
Electron affinity, χ (eV)	4.01	4.07	4.13
Refractive index (n_s) at $\lambda = 600$ nm	3.97	3.94	5.75
$B = \sqrt{N_c N_v}$ (Eq. (3.5))	5.23×10^{15}	1.66×10^{15}	2.10×10^{14}
Intrinsic carrier concentration, n_i (cm^{-3})	1.5×10^{10}	1.8×10^{6}	2.4×10^{13}
Electron mobility, μ_n cm^2/(V · s)	1350	8500	3900
Hole mobility, μ_p cm^2/(V · s)	480	400	1900

Table B.4 Useful SI prefix conversions.

Prefix	Symbol	Power of 10	Example
tera	T	10^{12}	1 TW = 10^{12} W
giga	G	10^{9}	1 GW = 10^{9} W
mega	M	10^{6}	1 MW = 10^{6} W
kilo	k	10^{3}	1 kW = 10^{3} W
(base unit)	–	10^{0}	1 W
milli	m	10^{-3}	1 mW = 10^{-3} W
micro	μ	10^{-6}	1 μm = 10^{-6} m
nano	n	10^{-9}	1 nm = 10^{-9} m
pico	p	10^{-12}	1 pF = 10^{-12} F

Appendix C

Key Charts and Graphs

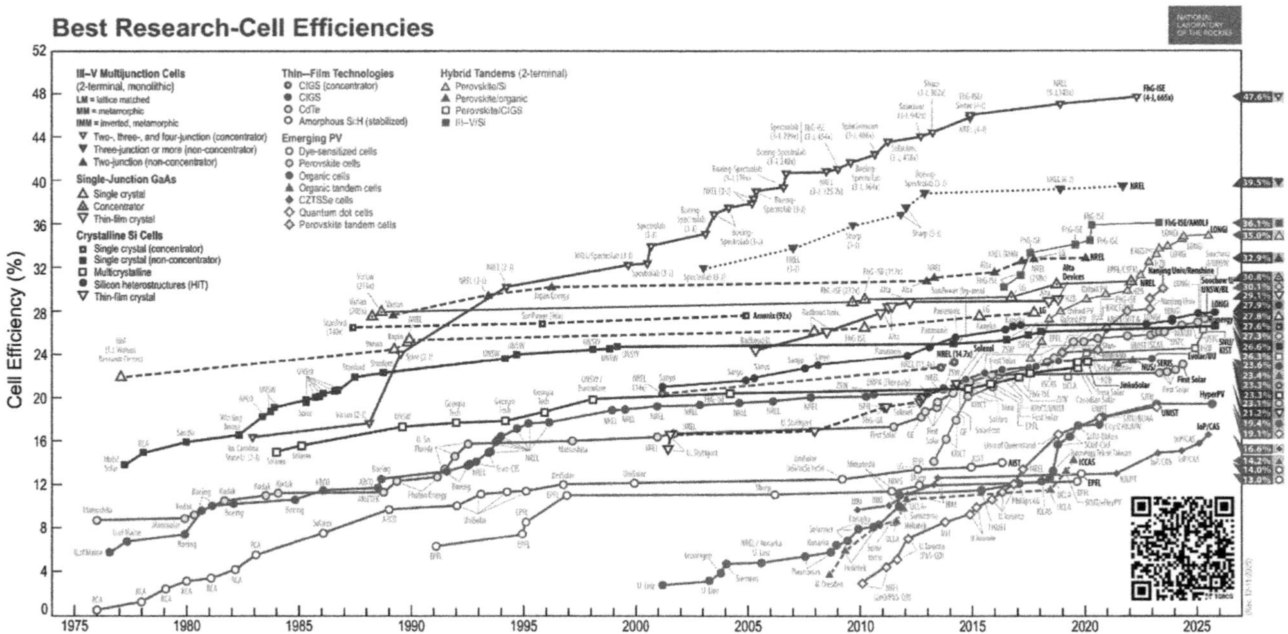

Figure 1.8 Best Research Solar Cell Efficiencies (1975–2025). *Source:* [3] / National Laboratory of the Rockies / Public domain. Scan QR code to access most recent solar cell efficiency tab [1].

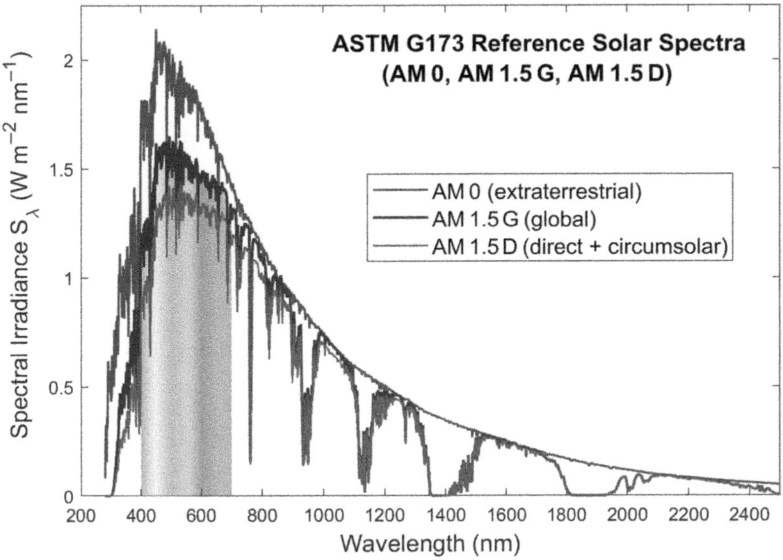

Figure 2.2 Reference solar spectral irradiance for extraterrestrial conditions (AM0), standard global terrestrial conditions (AM1.5G), and direct plus circumsolar terrestrial conditions (AM 1.5D), plotted as a function of wavelength. The shaded region indicates the visible spectral range (400–700 nm). Data are based on the ASTM G173 standard reference spectra, generated using the SMARTS radiative transfer model and distributed by the National Laboratory of the Rockies (NLR). *Source:* Adapted from [1].

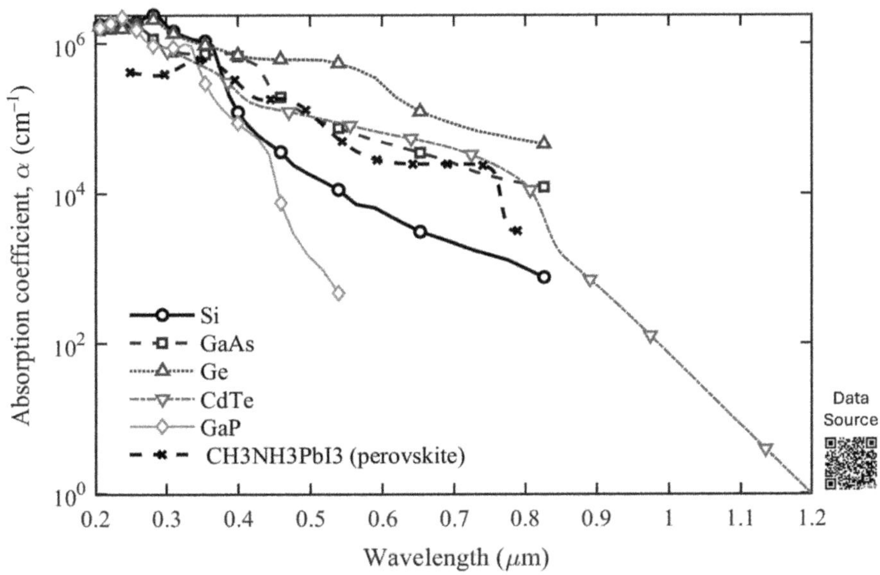

Figure 2.5 Absorption coefficients vs. wavelengths for materials often used in optical devices. *Source:* Adapted from [2].

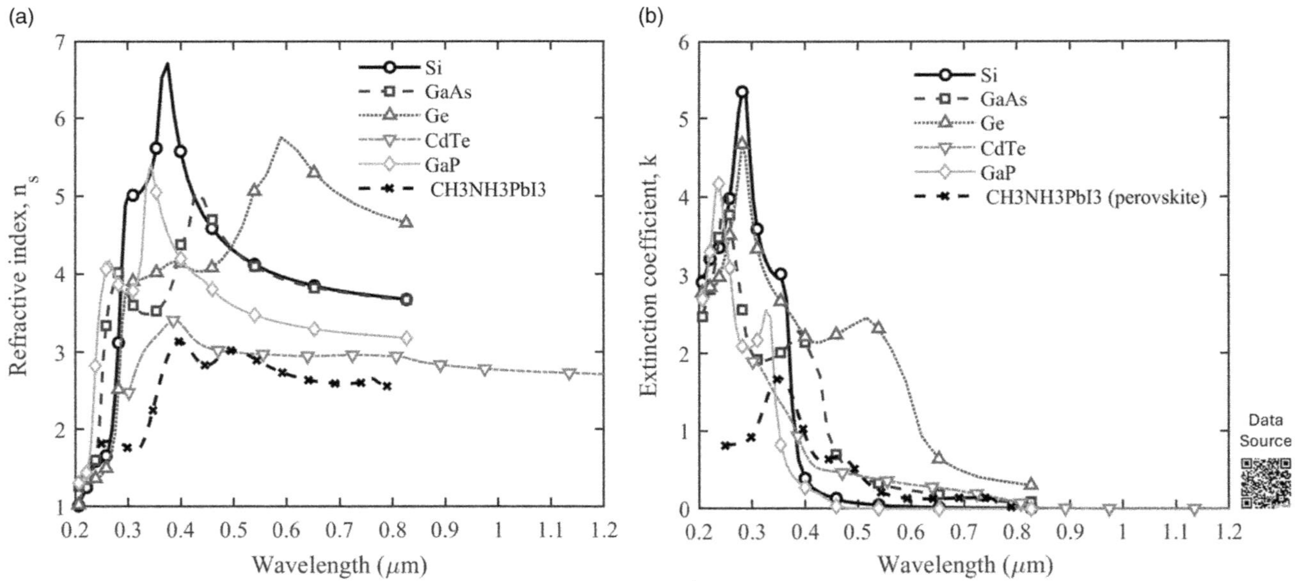

Figure 2.4 Optical properties of materials used in optical devices: (a) refractive index n_s and (b) extinction coefficient, k, as a function of wavelength. *Source:* Adapted from [2].

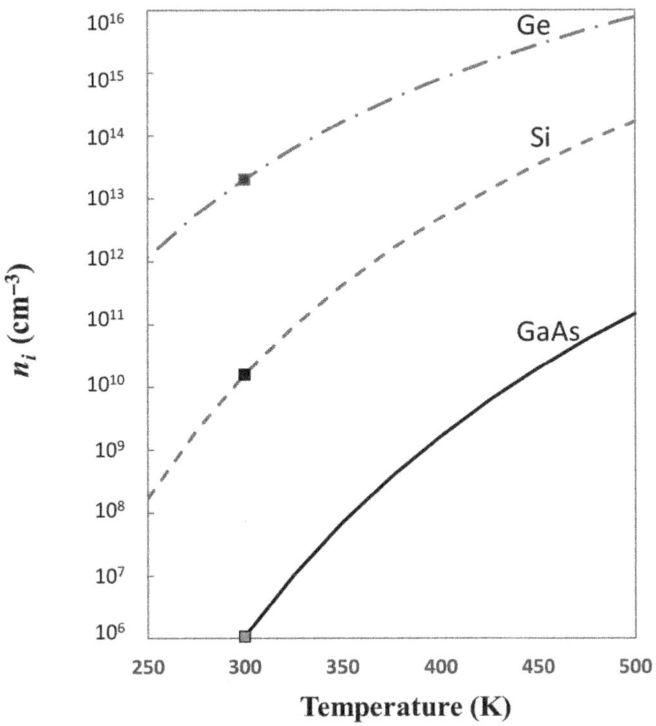

Figure 3.5 n_i as a function of temperature T (K) for Ge, Si, and GaAs.

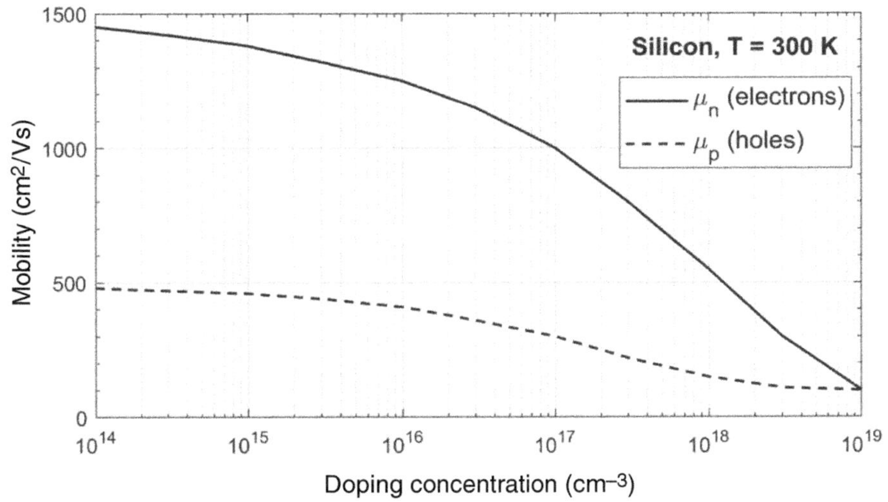

Figure 3.16 Electron and hole mobility vs. impurity concentration in silicon.

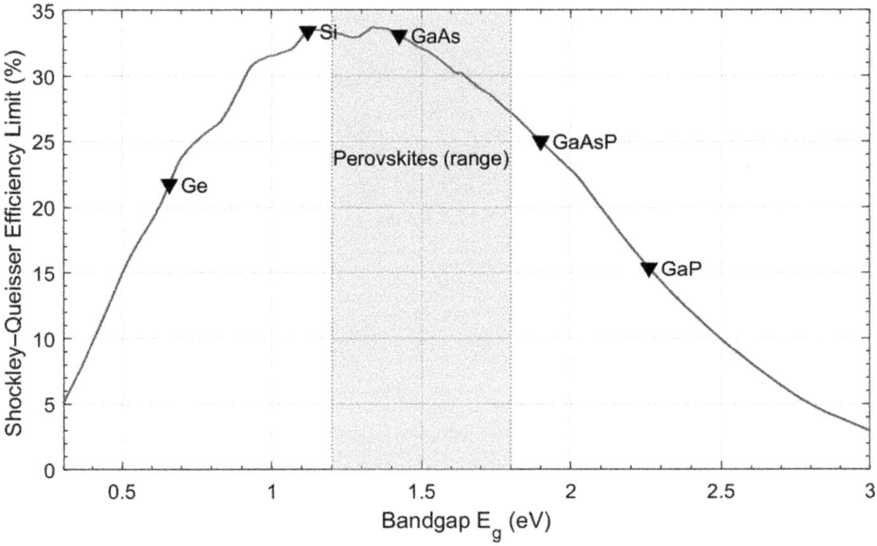

Figure 5.8 Efficiency vs. bandgap (Shockley–Queisser limit).

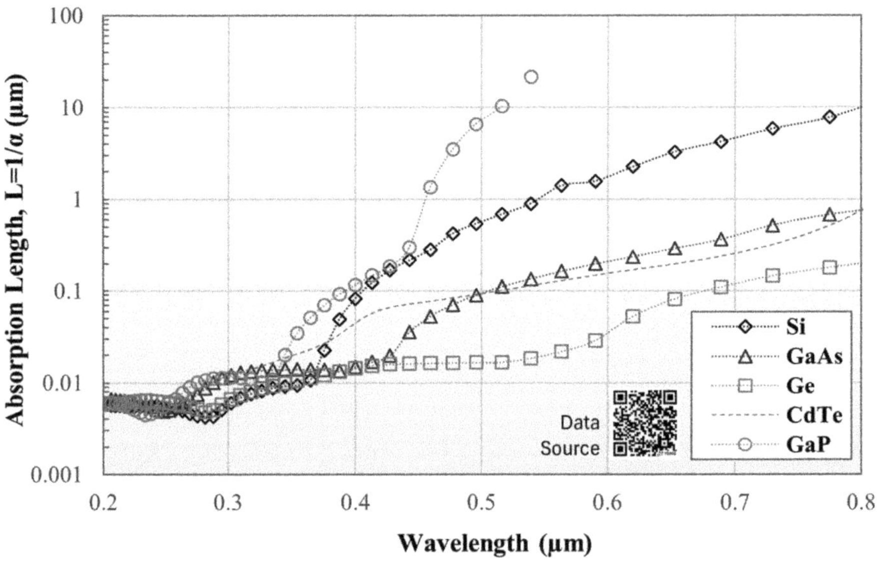

Figure 6.3 Absorption length $L = 1/\alpha$ (μm) as a function of wavelength (μm) for Si, GaAs, Ge, CdTe, and GaP materials. *Source:* Adapted from [2].

References

1 ASTM G173-03, "Reference solar spectral irradiance: air mass 0, air mass 1.5 global, and air mass 1.5 direct. ASTM International. Data distributed by the National Laboratory of The Rockies (NRL)," 2025. [Online]. Available: https://www.nlr.gov/grid/solar-resource/spectra-am1.5. [Accessed 2025].

2 refractiveindex.info, "Refractive index database: optical properties of inorganic materials," 2026. [Online]. Available: https://refractiveindex.info/?shelf=main&book=MoS2&page=Islam-o. [Accessed 2026].

3 National Laboratory of the Rockies, "Interactive best research-cell efficiency chart," 2025. [Online]. Available: https://www.nlr.gov/pv/cell-efficiency. [Accessed 2025].

Glossary

Absorber Layer The heart of the solar cell, where most of the photons are absorbed and the majority of the electron–hole pairs are created. It determines how many carriers are generated and how much power the solar cell can produce.

Absorption (*A*) Fraction of incident optical power absorbed in the material or multilayer stack.

Absorption Coefficient (α) Parameter describing optical attenuation inside a material, where intensity decays approximately as $I(x) = I_0 e^{-\alpha x}$.

Absorption Length A characteristic depth scale over which significant absorption occurs, defined as the inverse of the absorption coefficient. At one absorption length, approximately 63% of the incident light is absorbed.

Acceptor A dopant atom that creates a hole in the valence band, producing a *p*-type semiconductor.

Air Mass (AM) Standard measure of the optical path length sunlight travels through the atmosphere relative to the shortest path.

AM0 Spectrum Reference solar spectrum outside Earth's atmosphere.

AM1.5 Spectrum Standard terrestrial solar spectrum corresponding to air mass 1.5, used for PV performance comparisons.

Anti-Reflection Coating (ARC) A thin layer deposited on the surface of a solar cell to reduce reflection losses by causing destructive interference of reflected waves, increasing photon transmission into the absorber layer.

Bandgap The energy separation between the valence band and the conduction band in a semiconductor. It represents the minimum photon energy required to generate electron–hole pairs and fundamentally determines the electrical and optical properties of a material. In solar cells, the bandgap sets the trade-off between open-circuit voltage and short-circuit current and limits the maximum achievable efficiency.

Beer–Lambert Law A law describing the exponential decay of optical intensity as light propagates through an absorbing material, governed by the absorption coefficient and optical path length. Exponential decay of intensity in an absorbing medium: $\mathbb{I}(z) = \mathbb{I}e^{-\alpha x}$.

Built-in Electric Field The electric field established in the depletion region of a *pn* junction due to fixed ionized dopant atoms, which separates photogenerated charge carriers and enables their collection before recombination.

Built-in Potential The potential barrier formed across a *pn* junction at equilibrium that opposes carrier diffusion.

Capacity Factor The ratio of actual electricity generated by a power system over a given period to the maximum possible output if it operated continuously at full capacity.

Carbon Dioxide (CO_2) The primary greenhouse gas emitted through human activities, mainly from the combustion of fossil fuels such as coal, oil, and natural gas.

Carrier Mobility A measure of how easily electrons or holes move through a crystal under an electric field.

Climate Change Long-term changes in global and regional climate patterns, largely driven by the increased greenhouse effect caused by human activities.

Complex Refractive Index $\tilde{n}(\lambda)$ Optical index including absorption: $\tilde{n}(\lambda) = n_s(\lambda) - j\kappa(\lambda)$.

Concentrated Photovoltaics (CPV) Photovoltaic systems that use optical components such as lenses or mirrors to focus sunlight onto a small-area solar cell, increasing current density and power density.

Conduction Band The energy band in which electrons are free to move through the crystal and contribute to electrical conduction.

Conservation of Energy (Optics) Power balance relation: $R(\lambda) + \mathscr{T}(\lambda) + A(\lambda) = 1$.

Crystalline Silicon (c-Si) The dominant photovoltaic technology, including monocrystalline and polycrystalline silicon solar cells, widely used due to high efficiency, stability, and a mature manufacturing ecosystem.

Current Matching A design requirement in series-connected tandem solar cells where the total current density is limited by the subcell generating the smallest current.

Depletion Approximation The assumption that only fixed ionized impurities exist in the depletion region and mobile carriers are swept out.

Depletion Region The region near the *pn* junction depleted of mobile electrons and holes and containing fixed ionized dopant atoms.

Diffusion The motion of charge carriers from regions of high concentration to regions of low concentration.

Diffusion Current The current caused by the motion of carriers due to a concentration gradient.

Diffusion Length The distance that generated carriers can travel before recombining, influencing how effectively carriers reach the junction.

Direct-Bandgap Semiconductors Materials such as GaAs and CdTe in which optical transitions occur without a change in momentum. As a result, they exhibit strong absorption and short absorption lengths near the bandgap, allowing efficient solar cells to be fabricated using thin absorber layers.

Donor A dopant atom that donates an extra electron to the conduction band, creating an *n*-type semiconductor.

Doping The process of adding impurity atoms to a semiconductor crystal to increase the number of electrons or holes.

Drift The motion of charge carriers under the influence of an electric field.

Drift Current The current produced by the directed motion of electrons and holes in an electric field.

Efficiency, η The ratio of the electrical power output at the maximum power point divided by the input power from the Sun.

Einstein Relation The relationship between diffusion coefficient and mobility, linking diffusion and drift processes.

Electron–Hole Pair (EHP) A pair created when an electron is excited from the valence band to the conduction band, leaving behind an empty bond called a hole.

Emitter A heavily doped region that forms the *pn* junction and provides the built-in electric field needed to separate charge carriers. Its thickness and doping concentration affect recombination, conductivity, and absorption.

Energy Band Diagram A diagram showing the conduction band, valence band, and Fermi level across the *pn* junction.

Energy Bands Allowed energy levels formed when atoms come together to create a crystal, consisting of valence and conduction bands separated by a bandgap.

Energy Independence The ability of a country or region to meet its energy needs using domestically available energy resources, reducing reliance on imported fuels.

Equilibrium The condition where drift and diffusion currents cancel each other, resulting in zero net current.

Excess Carriers Additional electrons and holes generated beyond thermal equilibrium by optical absorption or electrical injection.

External Quantum Efficiency (*EQE*) Ratio of collected electrons to incident photons at each wavelength.

Extinction Coefficient $\kappa(\lambda)$ Imaginary part of the refractive index that quantifies absorption loss in a material.

Extrinsic Semiconductor A semiconductor whose electrical properties are modified by adding impurity atoms through doping.

Fermi–Dirac Statistics Quantum statistics used to calculate the probability that an electron occupies an energy state in a semiconductor.

Fermi Level The energy level at which the probability of electron occupancy is one half. Its position describes carrier concentrations in a semiconductor.

Fill Factor The area of the smaller rectangle ($I_m V_m$) divided by the area of the larger rectangle ($I_{sc} V_{oc}$) on the solar cell *IV* curve. It reflects how closely the curve resembles an ideal rectangle.

Finger Contacts Thin metal contacts used to collect charge carriers while minimizing shading losses. Their width and spacing balance series resistance and optical losses.

Forward Bias A bias condition where the applied voltage reduces the built-in potential and allows majority carrier diffusion across the junction.

Fossil Fuels Energy sources such as coal, oil, and natural gas formed from ancient organic matter, whose combustion releases greenhouse gases.

Frequency (ν) Number of wave cycles per second, related to wavelength through $c = \lambda\nu$.

Fresnel Coefficients Reflection and transmission coefficients at an interface, determined by refractive index mismatch.

Generation Rate, $\mathscr{R}$ The rate at which electron–hole pairs are created per unit volume per second.

Global Energy Transition The global shift from fossil fuel–based energy systems toward low-carbon and renewable energy sources.

Greenhouse Effect The phenomenon in which certain gases in Earth's atmosphere trap infrared radiation emitted by the Earth, leading to an increase in global temperature.

Greenhouse Gases (GHGs) Atmospheric gases, including carbon dioxide, methane, nitrous oxide, water vapor, and industrial gases, that absorb and re-emit infrared radiation, contributing to the greenhouse effect.

Hole The absence of an electron in the valence band, treated mathematically as a positively charged carrier.

Indirect Bandgap Semiconductor A material such as silicon that exhibits longer absorption lengths, especially near the band edge, requiring thicker absorber layers or light-trapping schemes.

Infrared Radiation (IR) A type of electromagnetic radiation emitted by objects with temperatures above absolute zero, invisible to the human eye but experienced as heat.

Installed Capacity The maximum amount of electrical power that an energy generation system can produce under ideal operating conditions.

Intensity ($\mathbb{I}$) Optical power per unit area carried by a wave.

Internal Quantum Efficiency (*IQE*) Ratio of collected electrons to absorbed photons, excluding reflection losses.

Intrinsic Carrier Concentration The number of electrons and holes present in a semiconductor at thermal equilibrium, determined by the bandgap and temperature.

Intrinsic Semiconductor A pure semiconductor with no external impurities added, where the concentration of electrons equals the concentration of holes.

Ionized Dopant Atoms Fixed charged atoms left behind when electrons or holes diffuse away from doped regions.

Levelized Cost of Electricity (LCOE) A measure of the average cost of electricity generation over the lifetime of a power system, including installation, operation, and maintenance costs.

Light Management Techniques such as anti-reflection coatings, surface texturing, and nanostructures that increase the number of photons absorbed by enhancing the optical path length.

Maximum Power Point The point on the solar cell IV curve where the product of current and voltage is maximum.

Metal Shading Loss Optical loss caused by metal contacts blocking incident light from reaching the absorber layer.

Methane (CH_4) A greenhouse gas emitted primarily from agricultural activities, livestock, rice cultivation, and waste management systems.

Minority Carrier Lifetime The average time an excess minority carrier exists before recombining.

Minority Carriers Charge carriers present in smaller concentration than the majority carriers in a doped semiconductor.

Multi-Junction Solar Cells High-efficiency photovoltaic devices consisting of multiple semiconductor junctions stacked in series, each designed to absorb a different portion of the solar spectrum.

Nitrous Oxide (N_2O) A greenhouse gas mainly released from agricultural fertilizer use in certain industrial processes.

***n*-Type Semiconductor** A semiconductor in which electrons are the majority carriers.

Open-Circuit Voltage, V_{oc} The voltage across solar cell when the current is zero.

Optical Loss Loss of usable photons due to reflection, parasitic absorption, or shading, reducing the number of photons available for carrier generation.

Optical Path Length (OPL) The effective distance that light travels inside a medium, accounting for both the physical thickness of the material and its refractive index.

Organic Solar Cell (OSC) A solar cell that uses organic semiconductors and can be fabricated on flexible substrates using solution-based processing.

Perovskite Solar Cells An emerging photovoltaic technology based on metal-halide perovskite materials, often hybrid organic–inorganic in composition, characterized by very high absorption coefficients and tunable bandgaps. These properties enable ultrathin absorber layers, rapid efficiency improvements, and strong potential for high-performance tandem solar cell applications.

Photocurrent Density (J_{ph}) Current density generated under illumination from collected photocarriers.

Photodiode A *pn* junction device in which absorbed photons change the current.

Photogenerated Current The current generated due to photon absorption and electron–hole pair generation.

Photon Energy (E) Energy carried by a photon: $E = h\nu = \frac{hc}{\lambda}$.

Photon Flux $\Phi(\lambda)$ Number of incident photons per unit area per unit time per wavelength interval.

Photovoltaic (PV) Effect The physical process by which semiconductor materials absorb photons from sunlight and release electrons, producing an electric current.

Photovoltaic (PV) Solar Cells Semiconductor devices, typically made of silicon, that convert sunlight directly into electricity using the photovoltaic effect.

Planck's Constant (h) Fundamental constant relating photon energy to frequency.

***pn* Junction** A junction formed between *p*-type and *n*-type semiconductor regions that creates the electric field required for carrier separation in a solar cell.

Poisson's Equation An equation relating the spatial variation of the electric field to the charge density in the junction.

Propagation Matrix Transfer matrix component describing phase accumulation and absorption within a layer.

***p*-Type Absorber** A lightly doped region where electrons are minority carriers and where most photons are absorbed and electron–hole pairs are generated for efficient collection.

***p*-Type Semiconductor** A semiconductor in which holes are the majority carriers.

Quantum Dot Solar Cell A solar cell that uses semiconductor nanocrystals with size-tunable bandgaps to absorb specific regions of the solar spectrum.

Quarter-Wave Thickness Film thickness giving a $\pi/2$ phase shift: $d = \frac{\lambda_0}{4n_s}$.

Quasi-Fermi Levels Separate energy levels used to describe electron and hole populations under nonequilibrium conditions.

Recombination Rate The rate at which electrons return to the valence band and recombine with holes.

Reflectance, $R(\lambda)$ Fraction of incident optical power reflected from a surface or multilayer stack.

Refractive Index, $n_s(\lambda)$ Real part of the complex refractive index of the material, determining the phase velocity of light and governing reflection, refraction, and optical path length.

Renewable Energy Energy derived from naturally replenished sources, such as solar energy, that produce little to no greenhouse gas emissions during operation.

Reverse Bias A bias condition where the applied voltage increases the built-in potential, widens the depletion region, and suppresses majority carrier diffusion across the junction.

Reverse Saturation Current The small current observed in a *pn* junction under reverse bias, arising from thermally generated minority carriers created near or within the depletion region of the junction.

Series Resistance Electrical resistance in the solar cell that limits current flow and reduces power output, influenced by contact design and material conductivity.

Shockley–Queisser Limit Maximum theoretical efficiency of a single-junction solar cell under standard illumination, derived using detailed balance between absorption and radiative recombination.

Short-Circuit Current Density, J_{SC} The current density when the voltage is zero in a solar cell.

Solar Energy Energy obtained from the Sun's radiation, which can be converted into electricity or heat using photovoltaic or solar thermal technologies.

Spectral Irradiance, $S_\lambda(\lambda)$ Optical power incident per unit area per unit wavelength, typically expressed in ($W\ m^{-2}\ nm^{-1}$).

Surface Texturing The use of micro- or nanoscale surface features to scatter and trap light inside a solar cell, increasing the probability of photon absorption.

Tandem Solar Cell A solar cell structure composed of two or more semiconductor junctions with different bandgaps, stacked to absorb different portions of the solar spectrum and improve energy conversion efficiency.

Thermal Equilibrium The state in which carrier generation, recombination, diffusion, and drift are balanced.

Thin-Film Solar Cells Solar cells fabricated using thin layers of semiconductor materials such as cadmium telluride or copper indium gallium selenide.

Transfer Matrix Method (TMM) Matrix method used to compute reflection and transmission through multilayer stacks by relating forward and backward wave amplitudes across interfaces and layers.

Transmittance, $\mathscr{T}(\lambda)$ Fraction of incident optical power transmitted through the layer or stack.

Transparent Conductive Oxide (TCO) A material that combines transparency and conductivity, used to improve carrier collection while allowing light to pass into the solar cell.

Two-Dimensional (2D) Materials Atomically thin semiconductor materials that exhibit strong light–matter interaction and are often used as charge-selective contacts or passivation layers.

Valence Band The energy band containing bound electrons involved in covalent bonding.

Yablonovitch Limit The theoretical upper limit on optical path length enhancement in a randomly scattering medium, defining the maximum achievable light trapping.

Index

Note: Page numbers in **bold** indicate the tables and *italic* indicate the figures.